Chemistry Research and Applications

Chemistry Research and Applications

The Future of Biorefineries
Waldemar Nyström (Editor)
2023. ISBN: 979-8-88697-524-6 (Hardcover)
2023. ISBN: 979-8-88697-528-4 (eBook)

Properties and Uses of Antimony
David J. Jenkins (Editor)
2022. ISBN: 979-8-88697-081-4 (Softcover)
2022. ISBN: 979-8-88697-088-3 (eBook)

The Science of Carbamates
Güllü Kaymak (Editor)
2022. ISBN: 978-1-68507-708-2 (Softcover)
2022. ISBN: 978-1-68507-872-0 (eBook)

Deep Eutectic Solvents: Properties, Applications and Toxicity
Carlos Eduardo de Araújo Padilha, PhD, Everaldo Silvino dos Santos, PhD, Francisco Canindé de Sousa Júnior, PhD, Nathália Saraiva Rios, PhD (Editors)
2022. ISBN: 978-1-68507-719-8 (Hardcover)
2022. ISBN: 978-1-68507-799-0 (eBook)

Polycyclic Aromatic Hydrocarbons: Sources, Exposure and Health Effects
Warren L. Gregoire (Editor)
2022. ISBN: 978-1-68507-626-9 (Softcover)
2022. ISBN: 978-1-68507-685-6 (eBook)

Cyanide: Occurrence, Applications and Toxicity
Bill M. Torres (Editor)
2022. ISBN: 978-1-68507-619-1 (Softcover)
2022. ISBN: 978-1-68507-670-2 (eBook)

More information about this series can be found at
https://novapublishers.com/product-category/series/chemistry-research-and-applications/

Catherine C. Bradley
Editor

What to Know about Lanthanum

Copyright © 2023 by Nova Science Publishers, Inc.
https://doi.org/10.52305/JWMC9723

All rights reserved. No part of this book may be reproduced, stored in a retrieval system or transmitted in any form or by any means: electronic, electrostatic, magnetic, tape, mechanical photocopying, recording or otherwise without the written permission of the Publisher.

We have partnered with Copyright Clearance Center to make it easy for you to obtain permissions to reuse content from this publication. Simply navigate to this publication's page on Nova's website and locate the "Get Permission" button below the title description. This button is linked directly to the title's permission page on copyright.com. Alternatively, you can visit copyright.com and search by title, ISBN, or ISSN.

For further questions about using the service on copyright.com, please contact:
Copyright Clearance Center
Phone: +1-(978) 750-8400   Fax: +1-(978) 750-4470   E-mail: info@copyright.com

NOTICE TO THE READER

The Publisher has taken reasonable care in the preparation of this book, but makes no expressed or implied warranty of any kind and assumes no responsibility for any errors or omissions. No liability is assumed for incidental or consequential damages in connection with or arising out of information contained in this book. The Publisher shall not be liable for any special, consequential, or exemplary damages resulting, in whole or in part, from the readers' use of, or reliance upon, this material. Any parts of this book based on government reports are so indicated and copyright is claimed for those parts to the extent applicable to compilations of such works.

Independent verification should be sought for any data, advice or recommendations contained in this book. In addition, no responsibility is assumed by the Publisher for any injury and/or damage to persons or property arising from any methods, products, instructions, ideas or otherwise contained in this publication.

This publication is designed to provide accurate and authoritative information with regard to the subject matter covered herein. It is sold with the clear understanding that the Publisher is not engaged in rendering legal or any other professional services. If legal or any other expert assistance is required, the services of a competent person should be sought. FROM A DECLARATION OF PARTICIPANTS JOINTLY ADOPTED BY A COMMITTEE OF THE AMERICAN BAR ASSOCIATION AND A COMMITTEE OF PUBLISHERS.

Additional color graphics may be available in the e-book version of this book.

Library of Congress Cataloging-in-Publication Data

ISBN: 979-8-88697-615-1

Published by Nova Science Publishers, Inc. † New York

Contents

Preface vii

Chapter 1   **Lanthanum: Complexes with Variable Coordination and Versatile Applications** 1
Vibha Vinayakumar Bhat and P. R. Chetana

Chapter 2   **The Effects of Lanthanum on Aquatic Organisms** 35
C. Figueiredo, P. Brito, M. Caetano and J. Raimundo

Chapter 3   **Synthesis and Properties of Nanodispersed Luminescent Structures Based on Lanthanum Fluoride and Phosphate for Optopharmacology and Photodynamic Therapy of Tumor Diseases Localized in Cranial Organs and Bone Tissues** 65
A. Kusyak, A. Petranovska, O. Oranska, S. Turanska, Ya. Shuba, D. Kravchuk, L. Kravchuk, G. Sotkis, V. Nazarenko, R. Kravchuk, V. Dubok, O. Bur'yanov, V. Chornyi, Yu. Sobolevs'kyy and P. Gorbyk

Chapter 4   **Removal and Recovery of Lanthanum from Aqueous Solutions by Biosorption** 95
Ellen C. Giese

Index 113

Preface

This book contains four chapters discussing essential facts about lanthanum. Chapter One describes lanthanum complexes with variable coordination and versatile applications. Chapter Two reviews the effects of lanthanum on aquatic organisms. Chapter Three discusses the synthesis and properties of nanodispersed luminescent structures based on lanthanum fluoride and phosphate for optopharmacology and photodynamic therapy of tumor diseases localized in cranial organs and bone tissues. Chapter Four discusses the removal and recovery of lanthanum from aqueous solutions by biosorption.

Chapter 1 - The rise of applications of lanthanum (III) metal ions in inorganic and bioinorganic fields are tremendous in recent years. The geometrical diversities due to variable coordination number, high Lewis acidity, and higher charge make La(III) ion bind to hard bases such as O, and N donor ligands making the complexes very beneficial in the field of luminescence, diagnostics and therapeutical agents. In recent years, rapid and continued interests have emerged in investigating and designing new molecules and metal complexes containing La(III) metal ion and other lanthanides in the field of medicine, MRI technology, immunoassays, separation techniques, etc. La(III)-complexes have been investigated for their applications as nucleic acid and protein binding agents, chemical nucleases, antiproliferative agents, antioxidants as well as antimicrobial agents. The stability induced by La(III) metal ion to the organic, bulkier ligands could be successfully utilized as a drug transport vehicle toward the target site in the biological systems. Here in this chapter, the authors would like to explain the synthetic methods of various La(III) complexes, their variable coordination geometry, applications in the field of luminescence, binding abilities towards nucleic acids and proteins, and cytotoxic activities against various cancerous cell lines, antimicrobial activities and antioxidant activities. The structural and application diversities of La(III)-complexes are graphically shown below.

Chapter 2 - Our high-technology society is increasingly reliant on technology-critical elements (TCE), such as the Rare Earth Elements (REE).

These elements are vital to "environmental-friendly" technologies, agriculture, and medical applications (e.g., MRI contrasts). With growing REE applications, emission to aquatic ecosystems rises through mining, ore processing, fossil fuel burn, and disposal of end life products (e.g., e-waste). Remarkably, maximum REE levels and discharges to the aquatic environment present no regulatory thresholds as evidence on the risks of increasing REE availability is limited. Nevertheless, REE are contaminants of emerging environmental concern that enter the aquatic medium through manifold ways. Lanthanum (La) is one of the most abundant and reactive REE in the environment and, thus, is relevant to understand its fate, bioavailability, bioaccumulation, and toxicological limits. This chapter analyses previously available scientific information on the La bioaccumulation, elimination, and ecotoxicology, focusing on freshwater and marine organisms. Lanthanum impacts biological responses and traits in multiple ways and may show interactive effects on marine biota coupled with other chemical and abiotic stressors. The authors will shortly explain the organisms' bioaccumulation and elimination capacity, while providing an overview of their hampering effects, at different levels of biological organization from microorganisms to plants, invertebrates, and vertebrates. Lanthanum can be detected in almost all biota although information on its speciation is regularly not available. Publications on trophic transfer of La was not found and hence biomagnification evidence is lacking. Nevertheless, La ecotoxicological outcomes are dispersed, mainly due to species specific differences and exposure conditions disparity that include exposure to distinct La compounds. This chapter highlights an urgent need for further studies on this emerging problem.

Chapter 3 - The aim of the work is the synthesis of nanodispersed phosphors based on lanthanum fluoride and lanthanum phosphate activated by terbium ($LaF_3:Tb^{3+}$ and $LaPO_4:Tb^{3+}$, respectively), promising for use in photodynamic therapy and optopharmacology, study of their structural properties and luminescence as well as the possibility of their use in nanocomposites (NC) with magnetically sensitive nanosized Fe_3O_4 carriers and 60S bioactive glass. Terbium-activated nanocrystalline lanthanum fluoride and lanthanum phosphate of hexagonal syngony were synthesized. Structural properties, chemical activity of surface, UV and X-ray luminescence spectra of the synthesized crystals have been studied. The possibility is shown to use them in NC with magnetically sensitive nanosized drug carriers and bioactive sol-gel glass. The acid-base nature of the active surface centers of LaF_3 and $LaF_3:Tb^{3+}$ NPs was determined. The parts of negatively α^-, positively α^+ charged and neutral α^0 active centers were calculated in the pH range of 2.4 – 12.7. The detected active centers of the surface can be represented by acidic (La^{3+}) and base (F^-) Lewis centers, as well as base Bronsted centers (OH^- groups). The obtained data are useful for optimization of the conditions of adsorption immobilization of molecules of photosensitive substances (photosensitizers) from physiological solution onto the surface of phosphors based on lanthanum fluoride. Ensembles of particles of magnetically sensitive NC $Fe_3O_4/LaF_3:Tb^{3+}$ of the core-shell type were synthesized. Conditions for the synthesis of NC did not significantly change the magnetic properties of their cores – the original single-domain Fe_3O_4 nanoparticles. 60S BG composites with nanodispersed crystalline $LaF_3:Tb^{3+}$ and $LaPO_4:Tb^{3+}$ in the dry state, and in distilled water, showed the presence of luminescence upon excitation by UV and X-rays. These data indicate the potential of research into nanodispersed phosphors based on lanthanum fluoride and lanthanum phosphate, their composites with magnetically sensitive nanosized carriers and bioactive glass, for use in optopharmacology and photodynamic therapy of tumor diseases localized in cranial organs and bone tissues. In addition, the results of research can be useful for technical applications, in particular, in the creation of luminescent detectors of high-energy electromagnetic radiation, the development of photo- and optoelectronic devices, etc

Chapter 4 - Biosorption is a cost-effective and simple technique for removing heavy metals and rare earth elements from an aqueous solution. Biosorption has also been considered a highly relevant green technology for replacing conventional unit operations of extractive metallurgy, viz. precipitation, liquid-liquid, solid-liquid extraction, and ion exchange. Biosorption is a physicochemical and metabolically-independent biological process based on various mechanisms, including absorption, adsorption, ion exchange, surface complexation, and precipitation, representing a biotechnological, cost-effective, innovative way for the recovery of lanthanum from aqueous solutions. This mini-review provides an overview and current scenario of biosorption technologies existing to recover lanthanum, seeking to address the possibilities of using a biotechnological approach for the recovery and separation of this high valued element in the REE production chain.

Chapter 1

Lanthanum: Complexes with Variable Coordination and Versatile Applications

Vibha Vinayakumar Bhat[1,*] and P. R. Chetana[2]

[1]Department of Chemistry, M S Ramaiah College of Arts, Science and Commerce, MSR Nagar, Bengaluru, Karnataka, India
[2]Department of Chemistry, Central College Campus, Bengaluru City University, Bengaluru, India

Abstract

The rise of applications of lanthanum (III) metal ions in inorganic and bioinorganic fields are tremendous in recent years. The geometrical diversities due to variable coordination number, high Lewis acidity, and higher charge make La(III) ion bind to hard bases such as O, and N donor ligands making the complexes very beneficial in the field of luminescence, diagnostics and therapeutical agents. In recent years, rapid and continued interests have emerged in investigating and designing new molecules and metal complexes containing La(III) metal ion and other lanthanides in the field of medicine, MRI technology, immunoassays, separation techniques, etc. La(III)-complexes have been investigated for their applications as nucleic acid and protein binding agents, chemical nucleases, antiproliferative agents, antioxidants as well as antimicrobial agents. The stability induced by La(III) metal ion to the organic, bulkier ligands could be successfully utilized as a drug transport vehicle toward the target site in the biological systems. Here in this chapter, we would like to explain the synthetic methods of various La(III) complexes, their variable coordination geometry, applications in the field of luminescence, binding abilities towards nucleic acids and proteins, and cytotoxic activities against various cancerous cell lines, antimicrobial

---

[*] Corresponding Author's Email: vibhamadhava@gmail.com.

In: What to Know about Lanthanum
Editor: Catherine C. Bradley
ISBN: 979-8-88697-615-1
© 2023 Nova Science Publishers, Inc.

activities and antioxidant activities. The structural and application diversities of La(III)-complexes are graphically shown below.

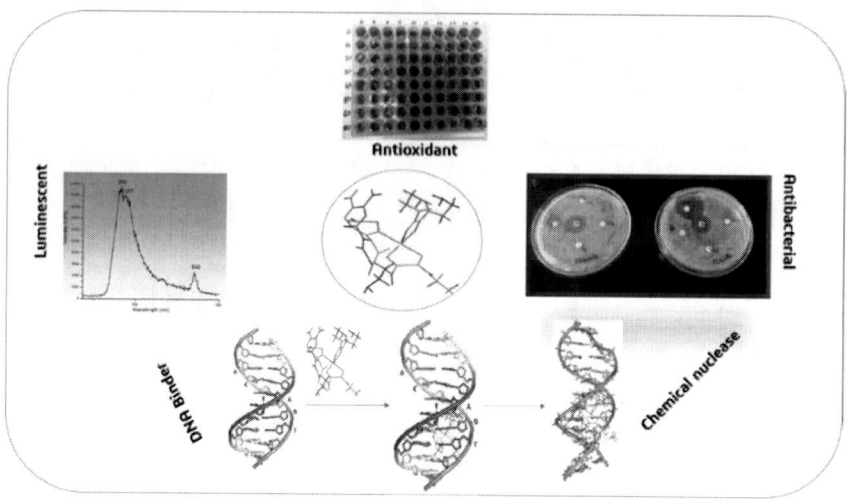

Keywords: Lanthanum(III) complex, DNA binder, chemical nuclease, antimicrobial, photoluminescent

Introduction

Lanthanum being an f-block element was discovered by Carl Gustav Mosander at the Karolinska Institute, Stockholm in 1839 while extracting cerium in an aqueous solution in its oxide form. It is well known that La(III) oxide is used in camera lenses, as an alloy in vehicle batteries, and in swimming pools to remove phosphates and algal growth [1, 2]. Tremendous work has been done in the field of bioinorganic chemistry employing transition metal complexes containing the various type of ligands. But metal complexes containing lanthanum (III) metal ion (La(III)) have been utilized in the pharmaceutical field, and diagnostics are quite rare. In recent years, rapid and continued interests have emerged in investigating and designing new molecules and metal complexes containing La(III) metal ions and other lanthanides in medicine, MRI technology, immunoassays, separation techniques, and chemical nucleases. To understand the chemistry of lanthanum complexes and their structure, knowing the history of lanthanum complexes is very important.

Literature Survey

After its first discovery, Axel Erdmann from the Karolinska Institute, Stockholm, 1839 extracted lanthanum from a new mineral [1]. These are the two existing historical backgrounds for lanthanum discovery. Later lanthanum is being investigated enormously for its structural diversity, geometry, coordination properties, and applications. The prominent application of La(III) or any other Ln(III) ions as MRI contrast agents is due to their ability to mimic biological cations Ca^{2+} and K^+ ions in their stable +3 oxidation state [3]. La(III) complexes could easily be synthesized either by template condensation or metastasis reactions. The coordination number of La(III) is as low as 3 to as high as 12 based on the types of ligands and their donor ability. The La-ligand complex formation could be identified by UV-Visible absorption and emission spectroscopy, FT-IR spectroscopy, NMR spectroscopy, ESI-Mass spectrometry, and X-ray diffraction methods. The detailed literature review is done under different subheadings where coordination numbers and various applications of La(III) complexes are discussed.

Coordination Chemistry of Lanthanum (III) Complexes

La(III) ion is oxophilic, and hard Lewis acid prefers hard bases such as O, and N donor atoms for effective coordination. The lower or higher coordination number is based on the coordinating ligands and solvents. As mentioned by Cotton et al. the crowding of ligands and their intermolecular interactions decide the coordination number of La(III) ion in a complex. The ligands such as halides, oxides, hydroxides, aqua as well as thiocyanate molecules influence La(III) ion to have a high coordination number whereas bulkier ligands such as tertiary alkyl groups, amide groups, silyl groups tend to make La(III) ion to attain low coordination numbers [4]. Herein, we have discussed various possible coordination numbers of the La(III) ion in complexes reported in literature.

Coordination Number 4
There are very few reports available with La(III) complexes having coordination number 4. The low coordination number for La(III) and other lanthanoid metals are quite unusual. Wang et al. have reported the synthesis and reactivity of a series of tris(boratabenzene) lanthanum complexes by

hydroboration and dehydrogenation coupling strategies. These complexes are known to be the first examples of complexes coordinated by three boratabenzene derivatives [5]. Zhao et al. have reported La(III) complexes of 1,10-phenanthroline derivatives. In this complex, the ligand 2-(1'-phenyl-2'-carboxyl-3'-aza-n-butyl)-1,10-phenanthroline contains soft nitrogen from phenanthroline group, hard nitrogen from amino group and carboxylate has hard oxygen atoms and all the three bases coordinated to La(III) ion to form 4 – coordinated complex [6]. The structure of the complex is as shown below:

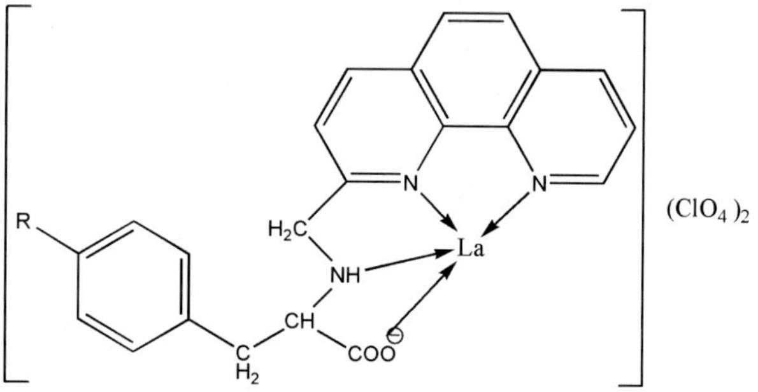

R = H (L1); OH (L2)

Figure 1. Structure of 4 – coordinate La(III) complex [6].

Earlier it was known that there would be the formation of adducts rather than the formation of complexes. The first stable four-coordinated adduct was found to be the La(III) and few other lanthanoid complexes of silylamide ligands and triphenylphosphine oxide. The La(III) and Lu(III) complexes were colourless whereas Eu(III) complex was yellow coloured [7].

Coordination Number 5

Though 5 – coordinate La(III) complexes are quite rare, few reports are available. Deacon et al. have investigated the 5 – coordinate La(III) complexes of organo-amide and aryloxy ligands. The geometry of the complex was distorted trigonal bipyramidal. The complex had shown two La – O bonds: phenol oxygen and tetrahydrofuran oxygen however both the bond lengths are different. Two phenolate oxygens having La-O bond lengths 2.25 and 2.234 Å occupy apical positions while THF oxygens occupy equatorial positions with La-O bond lengths 122.1, 124.8, and 96.5 Å respectively [8]. Clark et al. have reported the formation of distorted trigonal pyramidal complex of La(III) metal ion coordinated to four aryloxide and THF ligands. One aryloxide and

THF occupy axial positions. This is the example of sodium ate complex of La(III) organic complex [9].

Coordination Number 6

In lanthanum complexes, higher coordination numbers are very common. Complexes with central metal La(III) ion having coordination number is commonly observed with Schiff base ligands are reported. Siddappa et al. have reported the synthesis and antimicrobial activities of La(III) complex of (E)-3-((2-hydroxynaphthalen-1-yl)methyleneamino)-2-methylquinazoline-4(3H)-one. In this complex, La(III) ion exhibits octahedral geometry through the coordination of azomethine nitrogen and carboxylic oxygen with metal ion. The La(III) ion is 6-coordinated with three donor atoms from the Schiff base, two chloride ions and one O from water molecule [10]. La(III) complex of macrocyclic Schiff base containing oxamide groups have been reported by Aguairi et al. The complex was found to have coordination number 5 and 6 around the central La(III) ion [11]. Kulkarni et al. have reported the La(III) complexes of Schiff bases derived from coumarin derivatives [12]. Hegazy et al. have reported the La(III) complex of β – diketonates complexes with coordination number. Starynowicz et al. have reported the La(III) complexes of hexaazamacrocyclic imine with central La(III) ion having coordination number 6 [13]. In all the above-mentioned complexes, the central metal ion is coordinated to imine N (- HC = N) in common and anionic ligands.

6-coordinated La(III) complex was also reported by our group wherein [La^{III}(2PMO)(NO$_3$)$_3$](NO$_3$) acted as "complex as ligand" (PMO = N,N'-bis(2-pyridylmethyl)oxamide) [14]. Barta et al. have synthesized and investigated the La(III) and other Ln(III) complexes of 3-oxy-2-methyl-4-pyrone, 3-oxy-2-ethyl-4-pyrone, 3-oxy-1,2-dimethyl4-pyridinone and 3-oxy-2-methyl-4(1H)-pyridinone. In all these complexes, La(III) and other Ln(III) are coordinated with 3 equivalents of pyrone and pyridinone ligands respectively. These compounds were found to be beneficial in treating bone disorders [15]. Similar derivatives of pyridinone La(III) and other Ln(III) complexes were reported by Mawani et al. The ligands 3-hydroxy-2-methyl-1-(2-hydroxyethyl)-4-pyridinone, 3-hydroxy-2-methyl-1-(3-hydroxypropyl)-4-pyridinone, 3-hydroxy-2-methyl-1-(4-hydroxybutyl)-4-pyridinone, 3-hydroxy-2-methyl-1-(2-hydroxypropyl)-4-pyridinone, 3-hydroxy-2-methyl-1-(1-hydroxy-3-methylbutan-2-yl)-4-pyridinone, 3-hydroxy-2-methyl-1-(1-hydroxybutan-2-yl)-4-pyridinone, 1-carboxymethyl-3-hydroxy-2-methyl-4-pyridinone and 1-carboxyethyl-3-hydroxy-2-methyl4-pyridinone coordinate to La(III) and other Ln(III) ions in a tris ligand manner having coordination number 6 [16].

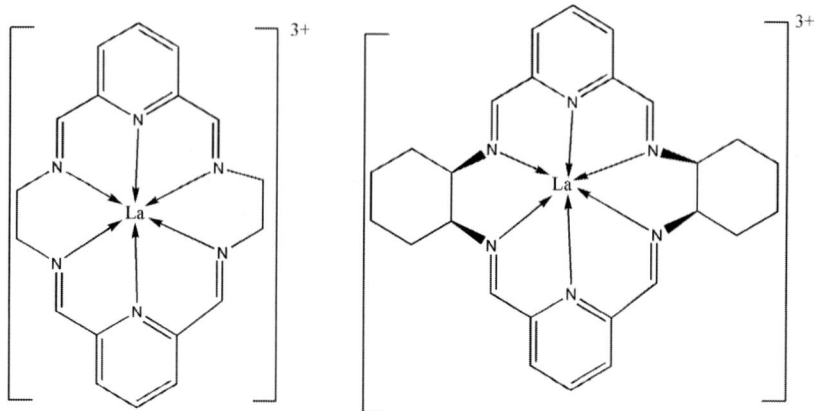

Figure 2. Structures of 6 - coordinated La(III) hexaazamacrocyclic Schiff base complexes [13].

Coordination Number 7

The higher coordination numbers of La(III) complexes will be achieved by sterically demanding bulkier ligands such as amides, oxamides, pyridine and so on. Fryzuk et al. have synthesized mixed ligand hydrocarbyl phosphine complexes of La(III), Y(III) and Lu(III). The complexes were found to be fluxional and thermally unstable and undergo cyclometallation reactions with continuous rearrangements [17]. Another example for bulkier ligands are 2,2'-bipyridyl and 1,10-phenanthroline. Roitershtein et al. have synthesized and characterized 2,2'-bipyridine,1,10-phenanthroline, and neocuproine complexes of La(III) halides. Herein they have reported the use of 2,2'-bipyrdyl ligand as anion using sodium amalgam as reducing agent. The complex [LaCl$_2$(Tp^{Me2})(bpy)] has 7-coordinated central La(III) ion with capped octahedral geometry [18]. Similar complexes were reported by Kretschmer et al. wherein the lanthanum dibromide complexes contained aminopyridinato and amidinate ligands. The complexes ApLaBr$_2$(THF)$_3$ and AmLaBr$_2$(THF)$_3$ contain sterically bulkier ligands Ap = (2,6-diisopropyl-phenyl)-[6-(2,4,6-triisopropyl-phenyl)-pyridin-2-yl]-amine and Am = N,N-bis-(2,6-diisopropylphenyl)-benzamidine respectively. The complexes were stable in solid state and the geometry around the central La(III) metal ion was found to be pentagonal bipyramid. The axial positions were occupied by two bromo ligands and three of the equatorial positions have been occupied by oxygen atoms of THF and two positions by N-atom of amide ligand [19]. Application of green chemistry principles has emerged in the synthesis of lanthanum complexes. Neelima et al. have synthesized La(III) complexes using 2,3–dihydro–1H–indolo–[2,3–b]–phenazin–4(5H)–ylidene) benzothiazole–2–amine and 3–(ethoxymethylene)–2,3–dihydro–1H–indolo–[2,3–b]–phenazin–4(5H)–one (Schiff base) ligands. In these complexes the La(III) ion was coordinated to N and S of Schiff base and three ionic chlorides making 7-coordinate environment around the central metal ion [20].

Coordination Numbers 8 - 12

Higher coordination numbers for La(III) can be from 8 - 12 depending on the size and nature of the ligands. Few of the examples have been discussed based on the reported complexes from the literature.

Coordination number 8 is commonly observed in La(III) and other Ln(III) complexes. Early in 1978, Karaghouli et al. have reported the crystal structure and geometry of the tetrakis lanthanum complex containing 2,2'-bipyridine dioxide. The complex [La(bpyO$_2$)$_4$] (ClO$_4$)$_3$ was nearly the first reported complex having La(III) in 8-coordination number and cubic geometry [21]. They have also reported the 8-coordinated La(III) complex having pyridine oxide ligand. The complex [La(PyO)$_8$]$^{3+}$ was reported to crystallize in *C2/c* monoclinic space group and had a coordination polyhedron nearly D_4 (422) symmetry. This indicated the geometry which was intermediate between a cube and a square antiprism [22]. The sterically crowded ligands generally act as chelates having hard base donor atoms. One such octadentate ligand was found to be 1,4,7,10-tetrakis(2-carbamoylethyl)-1,4,7,10-tetraazacyclododecane and its La(III) triflate complex was reported by Morrow et al. The ligand chelated La(III) through four oxygen atoms and four nitrogen atoms of the macrocyclic amide group. The La – O bonds were stronger compared to La – N bonds in the complex [23]. Majority of the lanthanum complexes having higher coordination numbers were perchlorate complexes. Zipp et al. have reported a series of disulphoxide complexes of La(III) wherein the coordination number around the central metal ion was 8, coordinated through the oxygen atoms of disulphoxide ligands [24]. The formation of eight coordinated La(III) complexes have been explored greatly. Lescop et al. have synthesized, characterized and studied the magnetic properties of La(III) complex series along with Gd(III) complexes. The ligand benzimidazole substituted nitronyl nitroxide acted as chelating ligand and formed two of the La(III) complexes as 8-coordinated with distorted dodecahedra and distorted cubic structure [25]. For some complexes, the coordination number 8 was satisfied by the solvent molecules along with the ligands. Example, diaquotris(isonicotinato) lanthanum(III) complex reported by Kay et al. This complex was reported to form a polymer in which metal ion is having four and two bridging carboxylate groups of isonicotinic acid and two aqua ions [26]. The solvation effects of water studied by density functional theory confirmed the 8-coordination number for La^{3+} [27].

Lanthanum (III) complexes having 9-coordinated La(III) ions are quite common. Tris La(III) complex having tridentate N-benzylbenzimidazolepyridine-2-carboxylate had 9-coordination number around the central La(III) ion having C_3 symmetry in solid state [28]. Similar examples can be seen for La(III) complexes having terpyridyl ligands [29], hydroxyquinoline carboxylate ligands [30], saccharinates [31] etc. Lanthanum chlorides also formed complexes with biomolecules such as sugars wherein they formed five Ln–O bonds from water molecules, three Ln–O bonds from hydroxyl groups of the D-ribopyranose, and one Ln–Cl bond from chloride ion [32].

As the steric crowding increases La(III) ion increases its size of the coordination sphere and can have 10-coordinated complexes. Examples are La(III) complexes with *N,N'*-donor bases such as 1,10-phenanthroline, dipyrido[3,2-d:2',3'-f]quinoxaline and dipyrido[3,2-a:2',3'-c]phenazine ligands [33], multidentate ethylenediamine-*N,N,N',N'*-tetraacetamide [34], tetraazamacycle [35], quinoline terpyridine [36] to name a few. One of the La(III) complex of 2-pyrazinecarboxylic acid having cationic hydrazinium ion is reported to have coordination number 10 [37]. In all the above mentioned complexes, La(III) ion was coordinated to hard bases O and N atoms, soft N atoms and the ions of their respective salts.

Carbohydrates are also known to form 11-coordinated complexes with La(III) ion. $La(NO_3)_3.6H_2O$ forms 11-coordinated complexes with neutral erythritol through two hydroxyl groups, six oxygen atoms from three nitrate ions and three water molecules. The erythritol ligand acted as bridging molecule between two La(III) ions in a binuclear complex [38]. The quinoline terpyridine ligand also formed 11-coordinate complex in tris ligand forms [36]. The bulkier ligands such as 1,4,7,10,13-pentaazacyclopenta-decane [39], β - dikenotes [40], tetradentate ONO donor hydrazones [41], *N*-substituted ethylcarbamate [42] and many others.

For La(III) though coordination number 12 is less known among other lanthanides, we can see few examples having ionic as well as coordination bonds. Among the series of lanthanide complexes of rare earth complexes of tributylammonium nitrate synthesized, the La(III) ion is coordinated twelve oxygens of six bidentate nitrate ligand. The complex was known to be a twist bistaggered polyhedral that crystallized in monoclinic space group [43]. Organometallic compounds can be coordinated to La(III) ion as ligands. Example, organotitanium fluoride [Hdmpy]$^+$[(C$_5$Me$_4$Et)$_2$Ti$_2$F$_7$]$^-$ when complexed with lanthanum(III) triflate, [La{(C$_5$Me$_4$Et)$_2$Ti$_2$F$_7$}$_3$] was formed wherein La(III) was coordinated to 12 fluorine atoms [44]. When 1,10-phenanthroline was a ligand, La(NO$_3$)$_3$.6H$_2$O formed anionic complex associated with protonated phenanthroline. The distorted icosahedral [La(NO$_3$)$_5$(phen)](phenH) complex crystallized in monoclinic space group P2$_1$/c. Since the complex was also associated with methanol solvent, there could be seen a weak interaction between protonated phenanthroline and methanol that gave two dimensional layered structure for the complex [45].

From the above information it can be observed that, depending on the donor atoms and available ligands, La(III) ion can have flexible coordination number, that can be as low as 4 and as high as 12. The electronic configuration of lanthanum metal is [Xe]5d^{1}6s^2. In stable +3 ionic state, the 5d and 6s orbitals will be vacant; expansion of coordination sphere is possible and hence La(III) ion can have variable coordination numbers. La(III) ion being hard Lewis acid, prefers hard bases such as ligands containing O and N donor atoms. Thus, La(III) ion can easily coordinate to ligands such as Schiff bases, β-diketonates, phenanthroline bases, acetates, thioacetates etc.

Binuclear Complexes

Lanthanum complexes have also exhibit variable coordination numbers in binuclear or polynuclear complexes. Few of the binuclear complexes have been generated when metal salt was added in excess during the routine synthesis of complexes. Example, when LaX$_3$.nH$_2$O (X = Cl, Br or NO3) when added to the to an aqueous ethanolic solution of 2,6-diacetylpyridine and carbohydrazide or thiocarbohydrazide resulted in binuclear La(III) complexes generated *in situ* [46]. Similarly, LaI$_3$(THF)$_3$ on treatment with excess of naphthalene (C$_{10}$H$_8$) in tetrahydrofuran (THF) results in the formation of binuclear complex. In this complex, there were two LaI$_3$(THF)$_3$ units bridged by the C$_{10}$H$_8$ unit. Each La(III) was a nine coordinated atom connected with η^4- naphthalene ligand at one corner and two trans iodide ions. In the same complex two types of La–O bonds have been identified. La–O which was *trans* to naphthalene was longer having 256.8 pm bond distance and shorter La–O having 251.1 pm which was *cis* to naphthalene [47]. By monitoring the reaction conditions, homonuclear as well as heterobinuclear complexes have been synthesized using macrocyclic Schiff bases formed by the reaction of 1,8-diamino-3,5-dioxaoctane and 4-chloro-2,6-diformylphenol. The complexes contained 6-coordinated La(III) metal centres in homobinuclear complexes and 6-coordinated La(III) and other Ln(III) ions in heterobinculear complexes [48]. When the metal ion with diversified coordination geometry and ligand having structural variation were mixed together, they were resulted in the formation of supramolecular materials. Example, $\{[La_2(na)_6(phen)_2][La_2(na)_6(phen)_2]\}$ and $[La_2(na)_6 (2,2\text{-bipy})_2]$ where na = 1-naphthoic acid, phen = 1,10-phenanthroline and bpy = 2,2'-bipyridine. Both the complexes had 9-coordinated La(III) ions but the ancillary ligands made the difference in their geometry. One complex had tricapped trigonal prism polyhedron whereas the other had distorted monocapped square antiprism La(III) ions [49]. Macrocyclic ligands having carboxymethyl groups which behave as pendant ligands resulted in binuclear complex. The binuclear La(III) ions were bridged between the macrocycle coordinated through O-atoms. Each La(III) ion had distorted tricapped trigonal prismatic coordination geometry in the complex [50].

Few ligands such as benzoyltrifluoroacetone and 4,4'-bipyridinedioxide coordinate with La(III) ions to form the three dimensional network of binuclear complexes [51]. La(III) ion also formed binuclear complexes with transition metals also. Ligands such as oxamides having N, O donor atoms form heterobinculear complexes. The substituted oxamide is a versatile coordinating ligand as it has three coordinating sites viz. N atom of amide, N atom of pyridyl ring and O atom of the amide carbonyl. Thus, it would be beneficial in synthesized binuclear metal complexes. The concept of "complex as ligands" is applicable in synthesizing binuclear complexes. Mononuclear aryl-substituted oxamide complex containing d-metal when used as ligand, the d-metal coordinates to N atoms based on the acid-base properties and hence leaves the carbonyl oxygen free for coordination of second metal ion preferably hard acids. Also, the combination of d-metal with f-metal in designing hetero-polynuclear complexes brings diverse range of metal complexes having potential applications in material science, bioinorganic and medicinal fields [52]. Copper oxamide complex and Nickel oxamide complex were utilized as complex as ligands in synthesizing heterobinuclear Cu(II)-La(III) complex and wherein 1,10-phenanthroline was used as ancillary ligand. The La(III) ion was found to be in 6-coordination environment wherein it is coordinated to two O-atoms of oxamide, two N-atoms of 1,10-phenanthroline and one bidentate nitrate ion [14].

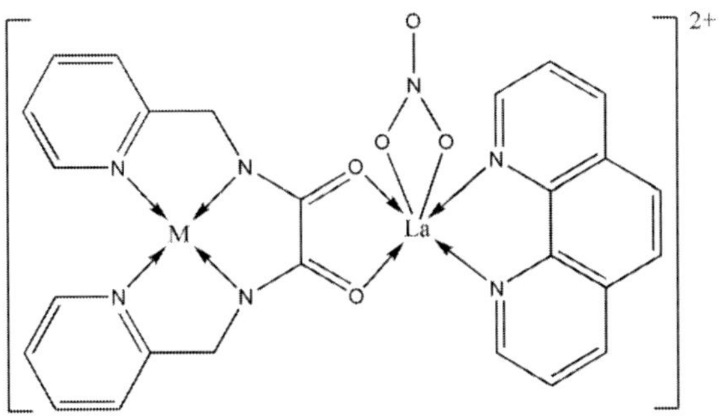

M = Cu(II), Ni(II)

Figure 3. Heterobinculear La(III) complex [14].

Applications of La(III) Complexes

As the diversified coordination numbers, La(III) complexes have found versatile applications in the field of medicinal field, luminescence, material science etc. The pharmaceutical properties of the metal complexes lie in their ability to bind to the biomolecules such as nucleic acids, proteins, enzymes and phospholipids. In this regard, many of the La(III) complexes have been evaluated for their biological properties both *in vitro* and *in vivo*. Few of the applications are discussed here.

La(III) Complexes as DNA Binders and Chemical Nucleases
Metal complexes which could interact with DNA very strongly and cleave it are highly beneficial in designing drug molecules. These are the prerequisites for a metal complex to possess anticancer and antimicrobial activities. Transition metal complexes are extensively studied for their binding interactions with biomolecules such as nucleic acids, proteins etc. But in recent years, rare earth metal complexes have been explored as binding probes, labeling agents, contrast agents in MRI and DNA foot printing agents because of their ability to bind to nucleic acids and proteins.

In 1997, Jonathan Hal et al. have synthesized a series of Ln(III) complexes and conjugated them into an oligonucleotides sequence. They have attached the complexes to oligonucleotides *via* covalent attachments using thiourea or amide linkages. These La(III) and other Ln(III) complexes are found to cleave RNA at 3′- end of the nucleotide [53]. Spectroscopic measurements can reveal the mode of binding of metal complexes to the biomolecules. La^{3+}-morin complex binds with DNA at the grooves [54]. Researchers have investigated several La(III) complexes that could bind and cleave nucleic acids. Example, La(III) and Ce(IV) complexes of 1,3-diamino-2-hydroxypropane-N,N,N',N'-tetraacetate (HPTA) ligand. It is found that La_2(HPTA) and Ce_2(HPTA) were the first complexes investigated to cleave DNA hydrolytically, forming significant number of linear forms of DNA [55]. Zhou et al. have synthesized rare earth complexes (La, Nd, Eu, Gd, Tb, Dy, Tm and Y) using quercetin (L) ligand and evaluated their antioxidative and antitumor activities. They have observed that $La(L)_3.6H_2O$ complex has shown higher binding affinity towards CT-DNA, superior antitumor activities against human liver cancer cells, than the ligand [56]. La(III) complexes of hypocrellin A have found to photocleave CT-DNA due to the longer triplet state of the complex in aerobic and anaerobic conditions [57]. Temperature dependent cleavage of plasmid DNA pBR322 and chromosomal DNA using lanthanide chlorides (Eu^{3+}, La^{3+}, Nd^{3+}, Pr^{3+} and Gd^{3+}) was studied in HEPES buffer (pH 7.0, 7.5 and 8.0). At elevated temperatures of 63 and 76°C, the complexes are found to cleave plasmid DNA into nick circular and linear form [58].

Ligands such as polyphenols, Schiff bases when coordinated to La(III) exhibit excellent DNA binding and cleavage activities. Example, La(III) complex of 7-methoxychromone-3-carbaldehyde benzoyl hydrazone Schiff Base binds to CT-DNA through intercalation and exhibit antioxidant activities against OH˙ and O_2^{-}˙ radicals [59]. The Schiff base derived from coumarin derivatives when complexed with La(III) also showed efficient cleavage of the plasmid DNA [12]. Polyphenols are good antioxidants. When catechin was coordinated with La(III) ion, it was found to bind CT-DNA through intercalative mode [60]. La(III) complex of N-(2-hydroxyacetophenyl)-2-aminopyrimidine Schiff base was found to cleave pUC 19 DNA efficiently [61]. Asadi et al. have synthesized two water-soluble mono-nuclear macrocyclic lanthanum(III) complexes of 2,6-diformyl-4-methylphenol with 1,3-diamino-2-propanol or 1,3-propylenediamine and studied their DNA binding, cleavage and cytotoxic activities. The DNA binding studies revealed the intercalation mode of binding. The anticancer experiments revealed that the hydrophilic complexes need a carrier to pass through the hydrophobic cell membrane. They have also synthesized two water-soluble La(III) hexaaza Schiff base complexes and studied their competitive binding affinity toward DNA and bovine serum albumin. The spectroscopic and docking studies revealed that complexes interacted with DNA through the minor groove. The binding studies with BSA revealed that the possible binding site was located on the vicinity of *Trp213* which was confirmed by docking simulations [62, 63].

Cytotoxic Activities of La(III) Complexes

The ability of the drug molecules or metal complexes to bind to tumor DNA and suppress their metabolism and growth can be termed as cytotoxic activity of the metal complexes. Various transition metal complexes can be found in literature that possess cytotoxic activity, but La(III) complexes which possess this ability are developed since very few years.

With emerging interests in lanthanide series complexes, Kostova et al. have worked enormously on the synthesis and investigation of anticancer or cytotoxic activities of La(III) and other Ln(III) complexes of various organic moieties. They synthesized La(III), Ce(III) and Nd(III) complexes using Warfaren and Coumachlor and screened their cytotoxic effects against carcinogenic lymphoma cells. The complexes found to exhibit higher cytotoxic activities than the ligands alone and salts. In continuation of their work, the same group has synthesized and tested the anticancer activities of La(III), Ce(III) and Nd(III) complexes of Niffcoumar against lymphoma derived P3HR1 cells. They have found that La(III) complexes are more active than Ce(III) and Nd(III) complexes [64] [65]. La(III), Ce(III) and Nd(III) complexes with 3,3′-benzylidenebis[4-hydroxycoumarin] found to be potent cytotoxic agents against were acute myeloid leukaemia derived HL 60 cell lines.. La(III) complexes of bis-coumarins i.e., 3,3′-benzylidene-bis(4-hydroxy-2H-1-benzopyran-2-one) and bis-(4-hydroxy-2-oxo-2H-chromen-3-yl)-(1H-pyrazol-3-yl)-methane showed *in vitro* cytotoxic activity against acute myeloid leukaemia-derived HL-60 and the chronic myeloid leukaemia-derived BV-173, reveals that the La(III) complexes exhibit cytotoxic activity probably via programmed cell death [66, 67]. La(III) complexes with 5-aminoorotic acid were screened for their *in vitro* antitumor activities against chronic myeloid leukemia derived K-562, over expressing the BCR-ABL fusion protein and the non-Hodgkin lymphoma derived DOHH-2 cell lines. They have observed that La(III) complexes exhibit *in vitro* cytotoxic effects in micromolar concentrations [68]. La(III), Ce(III) and Nd(III) complexes with coumarin-3-carboxylic acid exhibited *in vitro* cytotoxicity activities against acute myeloid leukemia-derived HL-60, the T-cell leukemia derived SKW-3 and the pre-B cell leukemia-derived Reh cell lines [69]. Cerium(III), lanthanum(III) and neodymium(III) complexes using 3,3′-(orthopyridinomethylene)di-[4-hydroxycoumarin] and evaluated their *in vitro* cytotoxic activities against HL-60 myeloid cells, which reveals that the complexes exhibit potent cytotoxic activities [70].

Researchers worldwide have been investigating La(III) complexes having various organic moieties for their cytotoxic activities. Zheng-Yin Yang et al. have synthesized La(III) complex with 1-phenyl-3-methyl-5-hydroxy-4-pyrazolyl phenyl ketone (PMBP)-isonicotinoyl hydrazone and discovered that La(III) complex possesses *in vitro* cytotoxic activity with inhibitory rate of 87.1% against Leukemia cells (L1210) [71]. Akhtar Hussain et al. have synthesized La(III) and Gd(III) complexes of pyrenyl-terpyridine ligands and investigated their photoactivated DNA cleavage and anticancer activity. 4'-phenyl-2,2':6',2''-terpyridine and 4'-(1-pyrenyl)-2,2':6',2''-terpyridine (py-tpy) are used as ligands. The py-tpy complexes are shown to cleave pUC19 DNA and exhibit remarkable photocytotoxicity in HeLa cells in UV-A light of 365 nm with apoptotic cell death having IC_{50} ~40 nM in light [72]. The mixed ligand complexes of La(III) and Gd(III) complexes containing 4'-phenyl-2,2':6',2''-terpyridine (ph-tpy), 4'-(1-pyrenyl)-2,2':6',2''-terpyridine (py-tpy) and diglucosylcurcumin ligands are found to be moderate binders to calf-thymus (CT) DNA. They cleaved plasmid supercoiled DNA into its nicked circular form in UV-A (365 nm) and visible light (454 nm) via $^1O_2$ and ·OH pathways. The complexes show remarkable photocytotoxic activity against HeLa cells in visible light (λ = 400 – 700 nm) [29].

Wang et al. synthesized La(III) complexes using (N,N'-bis-(1-carboxy-2-methylpropyl)-1,10-phenanthroline-2,9-dimethanamine), the L-valine derivative of 1,10-phenanthroline and investigated their *in vitro* cytotoxic activity against the HL-60 (the human leucocytoma) cells, HCT-8 (the human coloadenocarcinoma) cells, BGC-823 (the human carcinoma of stomach) cells, Bel-7402 (the human liver carcinoma) cells and KB (the human nasopharyngeal carcinoma) cells. The results revealed that La(III) complex was found to be highly active against all the cell lines and it has shown slightly higher activity than cisplatin against human nasopharyngeal carcinoma cells [73]. Biersack et al. synthesized La(III) complex and other transition metal complexes using natural melophlins and screened them for their cytotoxic activities against human A-498 kidney cancer cells. La(III) complex is found to exhibit antitumor activity at IC_{50} = 0.54 μM [74]. Durgo et al. synthesized La(III) complex using quercetin and investigated their cytotoxic and genotoxic effects on human cervical carcinoma cells. They have found that the complex exhibits cytotoxic activities in the concentration range 100 – 1000 mmol mL^{-1}. It was also observed that the same complex was found to induce dose-dependent pro-oxidative effects and cleaves single strand and double stranded DNA [75].

La(III) complexes having steric organic groups have shown profound anticancer activities. Chetana et al. synthesized several La(III) complexes having oxamides, phenanthroline bases as well as terpyridyl ligands and investigated their cytotoxic activities against human breast cancer cell lines (MCF 7) and human cervical carcinoma (HeLa) cell lines. The La(III) complexes exhibited potent antiproliferative activities against tested cell lines [14, 36].

Recently, researchers have began investigating the underlying mechanism of anticancer activities. [Tris(1,10-phenanthroline) lanthanum (III)]trithiocyanate complex was synthesized by Heffeter et al. and have been investigated for *in vitro* and *in vivo* anticancer activities against a human colon cancer xenograft. The apoptosis of tumor cells were induced by the La(III) complex as indicated by chromatin condensation, caspase substrate cleavage and mitochondrial membrane depolarization followed by cell cycle arrest [76]. New La(III) complex of cysteine was found to exhibit anticancer activities against MCF7 cell lines in time and dose dependent manners. The complex was also found to bind to BSA and βLG proteins that resulted in the conformational changes in the latter's structures [77]. La(III)-5-fluorouracil complexes were found to exhibit anticancer activities against Caco-2 cell lines and induced their apoptosis [78].

The above mentioned few investigations reveal that La(III) complexes can be beneficial in treating various cancerous cell lines. But further investigations and clinical trials are yet to be done for futuristic drug materials.

Antioxidant Activities of La(III) Complexes

La(III) complexes are found to exhibit good to excellent scavenging activity against hydroxyl and superoxide radicals. The reason may be due to the chelation of highly reactive hydroxyl or superoxide radicals to form stable complexes with La(III) ions. The complexes containing carbonyl, amine and hydroxyl groups exhibit good antioxidant activities. La(III) and Sm(III) complexes of 6-hydroxy chromone-3-carbaldehyde benzoyl hydrazone were prepared and tested for their in vitro antioxidant activity. The complexes showed significant radical quenching of hydroxyl and oxygen free radicals from the Fenton reaction [79]. La(III) complexes of polyphenols are known to exhibit significant antioxidant activities. Example, the La(III) complexes of quercetin [80] and catechin [60] quenched free radicals efficiently.

The ability of the metal complexes having excess donor atoms such as N or O that can chelate with Fe^{2+} ions or can donate H-atoms that can quench free radicals, is the prerequisite for any molecule or metal complex to behave as antioxidants. Heterobinuclear complexes of La(III) having oxamide and 1,10-phenanthroline ligands also behaved as antioxidants. The complexes could efficiently quench DPPH free radicals and could chelate to Fe^{2+} ions [14]. The La(III) and Eu(III) complexes of tridentate N-donor ligands N-(2-Aminoethyl)-1,3-propanediamine and 1-(2-Aminoethyl)piperazine have shown significant quenching of DPPH free radicals [81].

Recently, La(III) – perovskite nanoparticles have been synthesized and investigated for their biological activity. $La_{1-x}Ba_xMnO_3$ nanoparticles have quenched DPPH free radicals and also chelated Fe^{2+} ions [82].

The antioxidant activity of all the above mentioned complexes may be attributed to the fact that, the positive charge on the central La(III) ion increases the Lewis acidity and promotes the dissociation of hydrogen ions from the associated ligands. These protons can bind to the free radicals and thus can quench them effectively [81].

Antimicrobial Activities of La(III) Complexes

La(III) complexes have also been investigated for antimicrobial activities. Among these, Schiff bases play a major role. Avaji et al. have synthesized La(III) complexes with 3-substituted-4-amino-5-hydrazino-1,2,4-triazole Schiff bases and evaluated their biological activities. They have investigated that La(III) complex has shown high activity against *E. coli*. They have also synthesized La(III) complexes of thiocarbohydrazone and investigated their antimicrobial activity against bacteria (*E. coli* and *S. aureus*) and fungi (*A. Niger* and *A. Fumigates*). They have observed that La(III) complex is highly active against bacterial and fungal strains when compared to Schiff base alone [83, 84]. Gamel et al. have reported the synthesis and biological evaluation of Cu(II), Cr(II) and La(III) complexes of Schiff bases containing salen moiety. They have observed that, La(III) complex has shown higher antimicrobial activity against *E. coli*, *S. aureus* and *C. albicans* [85]. Siddappa et al. have synthesized the La(III) complex with Schiff base having quinazoline moiety and evaluated its antibacterial activity against methicillin-resistant *S. aureus* (MRSA). They have observed the higher antimicrobial activity of the La(III) complex than the Schiff base alone [10].

Andotra et al. have reported the synthesis and antimicrobial studies of La(III) tolyl/benzyldithiocarbonates. They have observed that antibacterial and antifungal activities of the complexes are more than the ligands. The

La(III) complexes containing heterocyclic ancillary ligands have shown potent activity against *K. pneumonia* and *B. cereus* [86]. Köse et al. have reported the synthesis of La(III) complex with Schiff base derived from 3-methoxy-2-hydroxybenzaldehyde and have studied against Gram positive and Gram negative bacterial strains and fungal strains. They have reported that coordination of oxygen atoms in the ligand with La(III) ion is the chief cause of their antimicrobial activity [87].

Alghool et al. have synthesized La(III) and few other Ln(III) metal complexes of Schiff base N-(2-hydroxybenzyl)-L-methionine acid and investigated their antibacterial against *E. coli, S. aureus* and antifungal activities against *A. flavus* and *C. albicans*. The complexes found to exhibit antimicrobial activities [88]. La(III) – phen complexes were synthesized by Niroomand et al. and investigated their biological activities. The [La(phen)$_3$Cl$_3$.OH$_2$] complex showed good antibacterial activity and also binding DNA at minor groove [89]. Recently, La(III) complexes have been synthesized based on mixed-ligand concept. Patil et al. have synthesized La(III) complexes of (2Z)-2-(N-hydroxyimino)-1,2-diphenylethan-1-ol as primary ligand and different amino acids as ancillary ligands. They have investigated the antibacterial activity of the new complexes. The complexes have shown effective antibacterial activity against positive strains than negative strains [90].

These reports reveal that La(III) ion when complexed with ligands of biological importance enhance their biological properties. Coordination of electron rich ligands such as Schiff bases (–CH = N), carbonyl compounds (–C = O), aromatic planar ligands cause the metal ion to share the positive charge with donor atoms of the ligands and thus polarity decreases and lipophilicity increases. Through chelation, the delocalization of electrons takes place throughout the ring and this favours the metal complex to penetrate through the biological membranes. Thus, biological efficiency of the ligands is enhanced on coordination [84, 91].

Photoluminescence Properties of La(III) Complexes

William DeW. Horrocks have observed the luminescence emission of dilute solutions of Eu(III) and Tb(III) ions when pulsed dye laser excitation source is used [92]. They have also reported the employment of luminescent properties of Eu(III) and Tb(III) ions in determining the number of coordinated water molecules to the Ln(III) metal ion that has already coordinated to the biomolecules such as proteins [93]. Based on these reports luminescent properties of La(III) complexes have been explored for various metal complexes. Waksrnan et al. have reported the synthesis and luminescent properties of crystal structures of the La(III), Eu(III) and Tb(III) cryptates of tris(bipyridine) macrobicyclic ligands [94]. Zhu et al. have synthesized two novel complexes containing Eu(III) and La(III) using trans-4-pyridylacrylate ligand and studied their luminescent properties in solid state. Of the two complexes, Eu(III) complex has shown strong red luminescence in solid state in room temperature and the lifetime was calculated to be 0.354 ± 0.001 ms [95]. Xiong et al. have reported the synthesis and luminescent properties of La complexes of phosphate hydrates. On doping with Ce^{3+}, these complexes exhibit intense luminescence under UV excitation [96]. Tang et al. have reported the preparation and luminescent properties of supramolecular lanthanide picrate complexes containing La, Pr, Nd, Eu, Gd, Tb and Er metal ions using the ligand 2,2'-[(1,2-phenylene)bis(oxy)]bis(N-benzylacetamide. It has been observed that the triplet energy level of the ligand matches with the excited level of Ln(III) ion and thus energy gets transferred from the ligand to the metal ion. Of all the complexes studied, Eu(III) complex has shown the strongest emission in solid state [97].

Dou et al. have synthesized La, Nd, Eu and Tb complexes using flexible salen type ligand and reported their luminescent properties. The effect of different solvents on the intensity of the luminescence is observed for the complexes. It has been shown that in solution state the luminescence has been quenched whereas the Tb complex in solid state has shown the strongest luminescence due to the shielding of Tb^{3+} ion by ligand molecule [98]. Galyametdinov et al. have reported the polarized luminescence properties of lanthanodimesogens i.e., liquid crystalline lanthanide complexes that include La, Nd, Eu and Yb of ligands β-diketonates and substituted 2,2'-bipyridine. They have reported that these complexes emit linearly polarized light in their nematic phase [99]. Regulacio et al. have discovered the room temperature luminescence of trivalent La, Pr, Sm, Eu, Gd, Tb and Dy complexes of dithiocarbamate and ancillary ligands 1,10-phenanthroline, 2,2'-bipyridine and 5-chloro-1,10-phenanthroline. The results reveal that Sm^{3+} and Pr^{3+} dithiocarbamate complexes exhibit more intensed luminescence than Eu^{3+}, Tb^{3+} and Dy^{3+} complexes. Whereas phosphorescence spectra of La^{3+} and Gd^{3+} analogous reveal the presence of lowest ligand localized triplet energy level [100].

Li et al. have reported the synthesis of Eu(III) β-diketonate complexes containing 1-(2-naphthoyl)-3,3,3-trifluoroacetonate and 2,2'-bipyridine and their co-luminescence effect when doped with La, Y, Gd and Tb ions. They have shown that the ligand transfers triplet state energy into emissive energy level of metal ions. This emission in the presence of other rare-earth ions on doping gets enhanced due to the intramolecular energy transfer between the non-fluorescing lanthanide ions and Eu(III) ions [101]. Roof et al. have reported the structural and photoluminescence properties of niobate complexes (LnKNaNbO$_5$) which contain Ln as La, Pr, Nd, Sm, Eu, Gd, and Tb metal ions. Of all the tested complexes, room temperature photoluminescence is exhibited by EuKNaNbO$_5$, GdKNaNbO$_5$ and TbKNaNbO$_5$ with characteristic bright orange-red, pink-purple and green emissions respectively. They have observed the influence of structure of niobate on the f - f and f - d transitions which are responsible for the photoluminescence properties [102].

Granadeiro et al. have reported the photoluminescent properties of Lanthanopolyoxometalates. They have studied organic-inorganic hybrid tungstate and molybdate materials of La, Ce, Sm, Eu, Tb, Er and Eu. Sensitization is observed in hybrid materials containing 3-hydroxypicolinate ligand, resulting in emission of Eu(III) complex in visible and near infrared regions. They have observed the enhancement of emission intensities on incorporation of picolinate ligands into the hybrid materials [103]. Hu et al. have reported the synthesis of La(III) and Sm(III) complexes of 2,2'-bipyridine-3,3'-dicarboxylic acid by hydrothermal method and studied their photo luminescent properties. The report reveals that, La(III) complex has shown broad blue emission peak at 428 nm in solid state at room temperature. This peak is attributed to the intra-ligand fluorescence, which when coordinates with La(III) ion, gets enhanced. This is due to the increased rigidity of the polymer on coordination with La(III) ion which in turn enhances the fluorescence [104]. Chai et al. have designed a series of lanthanide frameworks using *N,N'*-Diacetic acid imidazolium ligand. La(III) and Nd(III) complexes are synthesized using *cis-* and *trans-N,N'*-diacetic acid imidazolium. From their experimental results, they have confirmed the ligand-metal energy transfer in inducing luminescent properties to metal ions [105].

Yao et al. have reported the luminescence properties of lanthanide dimethyl sulfoxide compound solutions which contain Ce, La, Tb, Yb, Nd, Gd and Eu metal ions. They have classified Ln(III) complexes based on their spectral results. Accordingly, La(III)-DMSO complex belongs to type I, which has very broad excitation and emission spectra with shift in excitation and emission peaks to longer wavelengths [106]. Wang et al. have synthesized binuclear rare earth picrate complexes containing amide-based multidentate ligand, *N,N'*-1,2-ethanediyl-bis{2-[(*N,N*-diethylcarbamoyl)-methoxy]benzamide and investigated their photoluminescence properties. The metal ions include La, Nd, Eu, Gd, Tb, Dy, Yb and Y. They have reported that of all the complexes studied, Eu(III) complex has shown strong luminescence effect. They have also studied the effect of different solvents on the luminescence properties of Eu(III) complex. Eu(III) has shown good fluorescence intensities in ethyl acetate [107]. Bennett et al. have developed chiral lanthanide complexes using Y, La, Pr, Nd, Sm metal ions, bis(oxazolinylphenyl)amide (BOPA) ligand and alkyl and amide co-ligands and studied their photophysical properties using luminescence spectroscopy [108].

Mohanan et al. have synthesized a series of Schiff base complexes containing La(III), Pr(III), Nd(III), Sm(III), Gd(III), Dy(III) and Yb(III) ions and investigated the luminescence property of La(III)- Schiff base complex. The results reveal that the La(III) complex has shown increased intensity in its luminescence with characteristic blue shift around 410 nm. This band is attributed to the efficient transfer of energy from the triplet state of ligand to the lowest excited energy level of La(III). This reveals the importance of coordination between the Schiff base and the La(III) ion in enhancing the emissive properties [61].

Yin et al. have synthesized lanthanum cadmium complex containing 2,2'-bipyridine (bpy) ligand by hydrothermal method and described their photoluminescence properties in solid state. The results reveal that the complex exhibits intense emission band at 701 nm, is mainly because of the charge transfer by bpy ligand to metal ion. 2,2'-bipyiridine acts as antenna in sensitizing the La(III) ion [109]. Nedilko et al. have reported the luminescence properties of $LaBiVO_4$:MoEu. They have investigated that, on substituting Bi completely by La and partially Mo substitution for V, Eu^{3+} excitation spectra get extended upto 400 nm. Thus, it is reported that lanthanum and bismuth vanadates when doped with Eu^{3+}, the luminescent properties can be controlled [110].

Ln based metal ions are weak emitters as the f-electrons are shielded by $5p^66s^2$ subshells, they require a linker or chromophoric organic molecules to enhance the emission by transferring their energy. These linkers are generally organic molecules with conjugated aromatic system. When chromophores such as N,N-heterocyclic bases, Schiff bases, amides, oxamides, carboxylates etc. coordinate to Ln(III) ion, they function as UV-light collectors. These ligands absorb light in the UV-region, can sufficiently transfer their energy into the excited energy levels of lanthanide ions by a process called "antenna effect" through coordination. For the efficient transfer of energy, the ligand triplet energy states must closely match or slightly above the metal ion's resonance energy levels [111]. The unique physical properties of Ln(III) ions are greatly dependent on geometrical and molecular structures of their complexes. The overall quantum yields depend on the sensitivity of the *4f*-excited states to the coordinating chromophoric groups bearing N, O and C–H oscillators which suppress non-radiative deactivation and also enhance efficient energy transfer between the antenna and the coordinating Ln(III) ions [61, 112, 113].

Conclusion

The physical and chemical properties of a metal complex can be tuned by changing organic moieties of the ligands. This results in structural diversity, variation in coordination geometry and their applications. The above mentioned applications reveal that structure-activity-relationship can be derived based on structural variations of La(III) complexes that can be a boon in designing new molecules having multi-functional applications.

Conflict of Interest

Authors declare there are no conflict of interest regarding the work.

References

[1] *Periodic Table: History,* (n.d.). https://www.rsc.org/periodic-table/history (accessed May 7, 2022).
[2] *IYPT promotional pack,* Royal Society of Chemistry. (n.d.). https://www.rsc.org/iypt/iypt-elements/ (accessed May 7, 2022).
[3] Yerly, F., F.A. Dunand, E. Tóth, A. Figueirinha, Z. Kovács, A.D. Sherry, C.F.G.C. Geraldes, A.E. Merbach, Spectroscopic Study of the Hydration Equilibria and Water Exchange Dynamics of Lanthanide(III) Complexes of 1,7-Bis(carboxymethyl)-1,4,7,10-tetraazacyclododecane (DO2A), *European Journal of Inorganic Chemistry.* 2000 (2000) 1001–1006. https://doi.org/10.1002/(SICI)1099-0682(200005)2000:5<1001::AID-EJIC1001>3.0.CO;2-J.
[4] Cotton, S.A. Establishing coordination numbers for the lanthanides in simple complexes, *Comptes Rendus Chimie.* 8 (2005) 129–145. https://doi.org/10.1016/j.crci.2004.07.002.

[5] Wang, C., X. Leng, Y. Chen, Tris(boratabenzene) Lanthanum Complexes: Synthesis, Structure, and Reactivity, Organometallics. 34 (2015) 3216–3221. https://doi.org/10.1021/acs.organomet.5b00262.

[6] Zhao, G., F. Li, H. Lin, H. Lin, Synthesis, characterization and biological activity of complexes of lanthanum(III) with 2-(1'-phenyl- 2'-carboxyl-3'-aza-n-butyl)-1,10-phenanthroline and 2-(1'-p-phenol-2'-carboxyl-3'-aza-n-butyl)-1,10-phenanthroline, *Bioorganic & Medicinal Chemistry.* 15 (2007) 533–540. https://doi.org/10.1016/j.bmc.2006.09.032.

[7] Bradley, D.C., J.S. Ghotra, F.A. Hart, M.B. Hursthouse, P.R. Raithby, Low co-ordination numbers in lanthanoid and actinoid compounds. Part 2. Syntheses, properties, and crystal and molecular structures of triphenylphosphine oxide and peroxo-derivatives of [bis(trimethylsilyl)-amido]lanthanoids, *J. Chem. Soc.,* Dalton Trans. (1977) 1166–1172. https://doi.org/10.1039/DT9770001166.

[8] Deacon, G.B., B.M. Gatehouse, Q. Shen, G.N. Ward, E.R.T. Thiekink, Organoamido- and aryloxo-lanthanides—VII. The x-ray structure of five-coordinate [La(OC6H6Ph2-2,6)3(THF)2]·THF, *Polyhedron.* 12 (1993) 1289–1294. https://doi.org/10.1016/S0277-5387(00)84318-1.

[9] Clark, D.L., R.V. Hollis, B.L. Scott, J.G. Watkin, Alkali Metal Induced Structural Changes in Complexes Containing Anionic Lanthanum Aryloxide Moieties. X-ray Crystal Structures of (THF)La(OAr)2(μ-OAr)2Li(THF), (THF)La(OAr)2(μ-OAr)2Na(THF)2, and CsLa(OAr)4 (Ar = 2,6-i-Pr2C6H3), *Inorg. Chem.* 35 (1996) 667–674. https://doi.org/10.1021/ic9510472.

[10] Siddappa, K., S.B. Mane, D. Manikprabhu, La(III) complex involving the O,N-donor environment of quinazoline-4(3H)-one Schiff's base and their antimicrobial attributes against methicillin-resistant Staphylococcus aureus (MRSA), Spectrochimica Acta Part A: *Molecular and Biomolecular Spectroscopy.* 130 (2014) 634–638. https://doi.org/10.1016/j.saa.2014.03.115.

[11] Aguiari, A., S. Tamburini, P. Tomasin, P.A. Vigato, Lanthanide(III) macroacyclic and macrocyclic Schiff base complexes containing oxamidic groups, *Inorganica Chimica Acta.* 256 (1997) 199–210. https://doi.org/10.1016/S0020-1693(96)05441-2.

[12] Kulkarni, A., S.A. Patil, P.S. Badami, Synthesis, characterization, DNA cleavage and in vitro antimicrobial studies of La(III), Th(IV) and VO(IV) complexes with Schiff bases of coumarin derivatives, *European Journal of Medicinal Chemistry.* 44 (2009) 2904–2912. https://doi.org/10.1016/j.ejmech.2008.12.012.

[13] Starynowicz, P., J. Lisowski, Monomeric, dimeric and polymeric lanthanide(III) complexes of a hexaazamacrocyclic imine derived from 2,6-diformylpyridine and ethylenediamine, *Polyhedron.* 85 (2015) 232–238. https://doi.org/10.1016/j.poly.2014.08.032.

[14] Chetana, P., V. Bhat, M. Dhale, Hetero-Binuclear Complexes of Lanthanum (III) Using Bridging N,N'-Bis(2-Pyridylmethyl)Oxamide and Terminal 1,10-Phenanthroline: Synthesis, Characterization and Biological Evaluation, *International Journal of Pharmaceutical Sciences and Drug Research.* 10 (2018). https://doi.org/10.25004/IJPSDR.2018.100606.

[15] Barta, C.A., K. Sachs-Barrable, J. Jia, K.H. Thompson, K.M. Wasan, C. Orvig, Lanthanide containing compounds for therapeutic care in bone resorption disorders, *Dalton Trans.* (2007) 5019–5030. https://doi.org/10.1039/B705123A.

[16] Mawani, Y., J.F. Cawthray, S. Chang, K. Sachs-Barrable, D.M. Weekes, K.M. Wasan, C. Orvig, In vitro studies of lanthanide complexes for the treatment of osteoporosis, *Dalton Trans.* 42 (2013) 5999–6011. https://doi.org/10.1039/C2DT32373G.

[17] Fryzuk, M.D., T.S. Haddad, S.J. Rettig, Phosphine complexes of yttrium, lanthanum, and lutetium. Synthesis, thermolysis, and fluxional behavior of the hydrocarbyl derivatives MR[N(SiMe2CH2PMe2)2]2. *X-ray crystal structure of [cyclic] Y[N(SiMe2CHPMe2)(SiMe2CH2PMe2)][N(SiMe2CH2PMe2)2], Organometallics.* 10 (1991) 2026–2036. https://doi.org/10.1021/om00052a059.

[18] Roitershtein, D., Â. Domingos, L.C.J. Pereira, J.R. Ascenso, N. Marques, Coordination of 2,2'-Bipyridyl and 1,10-Phenanthroline to Yttrium and Lanthanum Complexes Based on a Scorpionate Ligand, *Inorg. Chem.* 42 (2003) 7666–7673. https://doi.org/10.1021/ic034838+.

[19] Kretschmer, W.P., A. Meetsma, B. Hessen, N.M. Scott, S. Qayyum, R. Kempe, Lanthanum Dibromide Complexes of Sterically Demanding Aminopyridinato and Amidinate Ligands, *Journal of inorganic and general chemistry.* 632 (2006) 1936–1938. https://doi.org/10.1002/zaac.200600142.

[20] Neelima, K. Poonia, S. Siddiqui, M. Arshad, D. Kumar, In vitro anticancer activities of Schiff base and its lanthanum complex, Spectrochimica Acta Part A: *Molecular and Biomolecular Spectroscopy.* 155 (2016) 146–154. https://doi.org/10.1016/j.saa.2015.10.015.

[21] Al-Karaghouli, A.R., R.O. Day, J.S. Wood, Crystal structure of tetrakis(2,2'-bipyridine dioxide)lanthanum perchlorate: an example of cubic eight-coordination, *Inorg. Chem.* 17 (1978) 3702–3706. https://doi.org/10.1021/ic50190a076.

[22] Al-Karaghouli, A.R., J.S. Wood, Crystal and molecular structures of the octakis(pyridine N-oxide) lanthanide complexes M(PyO)8(ClO4)3, M = lanthanum and neodymium, *Inorg. Chem.* 18 (1979) 1177–1184. https://doi.org/10.1021/ic50195a001.

[23] Morrow, J.R., S. Amin, C.H. Lake, M.R. Churchill, Synthesis, structure, and dynamic properties of the lanthanum(III) complex of 1,4,7,10-tetrakis(2-carbamoylethyl)-1,4,7,10-tetraazacyclododecane, ACS Publications. (2002). https://doi.org/10.1021/ic00073a017.

[24] Zipp, A.P., S.G. Zipp, Disulfoxide complexes with lanthanum perchlorate, *Journal of Inorganic and Nuclear Chemistry.* 42 (1980) 395–397. https://doi.org/10.1016/0022-1902(80)80013-3.

[25] Lescop, C., E. Belorizky, D. Luneau, P. Rey, Synthesis, Structures, and Magnetic Properties of a Series of Lanthanum(III) and Gadolinium(III) Complexes with Chelating Benzimidazole-Substituted Nitronyl Nitroxide Free Radicals. Evidence for Antiferromagnetic GdIII−Radical Interactions, *Inorg. Chem.* 41 (2002) 3375–3384. https://doi.org/10.1021/ic0200038.

[26] Kay, J., J.W. Moore, M.D. Glick, Structural studies of bridged lanthanide(III) complexes. Diaquotri(nicotinic acid)holmium(III) hexa(isothiocyanato)chromate

(III) dihydrate and diaquotris(isonicotinato)lanthanum(III), *Inorg. Chem.* 11 (1972) 2818–2827. https://doi.org/10.1021/ic50117a047.

[27] Shi, T., A.C. Hopkinson, K.W.M. Siu, Coordination of Triply Charged Lanthanum in the Gas Phase: Theory and Experiment, *Chemistry – A European Journal.* 13 (2007) 1142–1151. https://doi.org/10.1002/chem.200601074.

[28] Shavaleev, N.M., S.V. Eliseeva, R. Scopelliti, J.-C.G. Bünzli, Influence of Symmetry on the Luminescence and Radiative Lifetime of Nine-Coordinate Europium Complexes, *Inorg. Chem.* 54 (2015) 9166–9173. https://doi.org/10.1021/acs.inorgchem.5b01580.

[29] Hussain, A., K. Somyajit, B. Banik, S. Banerjee, G. Nagaraju, A.R. Chakravarty, Enhancing the photocytotoxic potential of curcumin on terpyridyl lanthanide(III) complex formation, *Dalton Trans.* 42 (2012) 182–195. https://doi.org/10.1039/C2DT32042H.

[30] Feng, R., F.-L. Jiang, M.-Y. Wu, L. Chen, C.-F. Yan, M.-C. Hong, Structures and Photoluminescent Properties of the Lanthanide Coordination Complexes with Hydroxyquinoline Carboxylate Ligands, *Crystal Growth & Design.* 10 (2010) 2306–2313. https://doi.org/10.1021/cg100026d.

[31] Vasilescu, I.M., M. Taş, P.C. Junk, Structural diversity of lanthanum saccharinates induced by 1,10-phenanthroline: A synthetic and X-ray crystallographic study, *Polyhedron.* 87 (2015) 259–267. https://doi.org/10.1016/j.poly.2014.11.009.

[32] Lu, Y., G. Deng, F. Miao, Z. Li, Metal-ion interactions with sugars. Crystal structures and FT-IR studies of the LaCl3–ribopyranose and CeCl3–ribopyranose complexes, *Carbohydrate Research.* 339 (2004) 1689–1696. https://doi.org/10.1016/j.carres.2004.04.009.

[33] Hussain, A., D. Lahiri, M.S. Ameerunisha Begum, S. Saha, R. Majumdar, R.R. Dighe, A.R. Chakravarty, Photocytotoxic Lanthanum(III) and Gadolinium(III) Complexes of Phenanthroline Bases Showing Light-Induced DNA Cleavage Activity, *Inorg. Chem.* 49 (2010) 4036–4045. https://doi.org/10.1021/ic901791f.

[34] Clapp, L.A., C.J. Siddons, J.R. Whitehead, D.G. VanDerveer, R.D. Rogers, S.T. Griffin, S.B. Jones, R.D. Hancock, Factors Controlling Metal-Ion Selectivity in the Binding Sites of Calcium-Binding Proteins. The Metal-Binding Properties of Amide Donors. *A Crystallographic and Thermodynamic Study, Inorg. Chem.* 44 (2005) 8495–8502. https://doi.org/10.1021/ic050632s.

[35] Jiang, B., M. Wang, C. Li, J. Xie, DNA-binding and hydrolytic cleavage promoted by tetraazamacycle La(III) and Ce(III) complexes, *Med Chem Res.* 22 (2013) 3398–3404. https://doi.org/10.1007/s00044-012-0357-7.

[36] Chetana, P.R., D.R. Navya, V.V. Bhat, B.S. Srinatha, M.A. Dhale, Studies on DNA Interactions and Biological Activities of Lanthanum(III) Complexes with 4-Quinoline Terpyridine and 1,10-Phenanthroline, *Asian Journal of Chemistry.* 31 (2019) 1265–1274.

[37] Premkumar, T., S. Govindarajan, N.P. Rath, V. Manivannan, New nine and ten coordinated complexes of lanthanides with bidentate 2-pyrazinecarboxylate containing hydrazinium cation, *Inorganica Chimica Acta.* 362 (2009) 2941–2946. https://doi.org/10.1016/j.ica.2008.12.014.

[38] Yang, L., X. Hua, J. Xue, Q. Pan, L. Yu, W. Li, Y. Xu, G. Zhao, L. Liu, K. Liu, J. Chen, J. Wu, Interactions between Metal Ions and Carbohydrates. Spectroscopic Characterization and the Topology Coordination Behavior of Erythritol with Trivalent Lanthanide Ions, *Inorg. Chem.* 51 (2012) 499–510. https://doi.org/10.1021/ic2019605.

[39] Li, X.D., Gan, M. Tan, X. Wang, Synthesis and characterization of lanthanide complexes of 1,4,7,10,13-pentaazacyclopentadecane, *Polyhedron.* 16 (1997) 3991–3995. https://doi.org/10.1016/S0277-5387(97)00181-2.

[40] Bray, D.J., J.K. Clegg, L.F. Lindoy, D. Schilter, Self-assembled Metallo-supramolecular Systems Incorporating β-Diketone Motifs as Structural Elements, in: R. van Eldik, K. Bowman-James (Eds.), *Advances in Inorganic Chemistry*, Academic Press, 2006: pp. 1–37. https://doi.org/10.1016/S0898-8838(06)59001-4.

[41] Patil, S., V. Naik, D. Bilehal, N. Mallur, Synthesis, Spectral and Antimicrobial Studies of Lanthanide (III) Nitrate Complexes with Terdentate ONO Donor Hydrazones, Journal of Experimental Sciences 2011, 2(7): 15-20. 2 (2011) 15–20.

[42] Casotto, G., B. Zarli, G. Faraglia, L. Sindellari, N-substituted ethylcarbamate complexes of thorium(IV) and lanthanum(III) nitrates, *Inorganica Chimica Acta.* 95 (1984) 247–250. https://doi.org/10.1016/S0020-1693(00)84892-6.

[43] Yan, C., Y. Zhang, S. Gao, B. Li, C. Huang, G. Xu, Synthesis and structure of tributylammonium hexanitrate lanthanum, *Journal of Alloys and Compounds.* 225 (1995) 385–389. https://doi.org/10.1016/0925-8388(94)07043-1.

[44] Perdih, F., A. Demšar, A. Pevec, S. Petriček, I. Leban, G. Giester, J. Sieler, H.W. Roesky, Synthesis and the crystal structures of a monoanionic tetrafluorodentate ligand and its complex with lanthanum ion, *Polyhedron.* 20 (2001) 1967–1971. https://doi.org/10.1016/S0277-5387(01)00789-6.

[45] Deng, Y.-H., G.-R. Gao, J. Liu, H. Liy. Synthesis and Crystal Structure of an Unusual Dodecacoordinated Lanthanum Complex by Phenanthroline and Nitrate (phenH)_2[La(NO_3)_5(phen)]·CH_3OH, *Chinese Journal of Structural Chemistry* 30 (2011) 690–696.

[46] Pandey, O.P. Mononuclear and binuclear lanthanum(III) complexes of macrocyclic ligands derived from 2,6-diacetylpyridine, *Polyhedron.* 6 (1987) 1021–1025. https://doi.org/10.1016/S0277-5387(00)80948-1.

[47] Fedushkin, I.L., M.N. Bochkarev, H. Schumann, L. Esser, Binuclear complexes of La(III) and Eu(II) with the bridging naphthalene dianion. Synthesis and X-ray crystallographic analysis of [μ2-η4:η4-C10H8][LaI2(THF)3]2 and [μ2-η4:η4-C10H8][EuI(DME)2]2, *Journal of Organometallic Chemistry.* 489 (1995) 145–151. https://doi.org/10.1016/0022-328X(94)05145-2.

[48] Guerriero, P., P.A. Vigato, J.-C.G. Bünzli, E. Moret, Macrocyclic complexes with lanthanoid salts part 38. Synthesis and luminescence study of homo- and hetero-binuclear complexes of lanthanides with a new cyclic compartmental Schiff base, *J. Chem. Soc., Dalton Trans.* (1990) 647–655. https://doi.org/10.1039/DT9900000647.

[49] Yang, E.-C., P.-X. Dai, X.-G. Wang, X.-J. Zhao, Two Discrete Lanthanum(III) Complexes with Bulky Aromatic Mixed Ligands: Syntheses, Crystal Structures and

Fluorescent Properties, *Journal of inorganic and general chemistry.* 635 (2009) 346–350. https://doi.org/10.1002/zaac.200800392.

[50] Inoue, M.B., M. Inoue, Q. Fernando, Yttrium(III) and lanthanum(III) complexes of a 15-membered macrocycle with three pendant carboxymethyl groups, *Acta Cryst* C. 50 (1994) 1037–1040. https://doi.org/10.1107/S0108270193011151.

[51] Ma, S.-L., W.-X. Zhu, G.-H. Huang, D.-Q. Yuan, X. Yan, Novel three-dimensional network of lanthanum(III) complex {[La2(BTA)6(4,4'-bpdo)1.5]·1.5H2O}n (BTA = benzoyltrifluoroacetone; 4,4'-bpdo=4,4'-bipyridine dioxide), *Journal of Molecular Structure.* 646 (2003) 89–94. https://doi.org/10.1016/S0022-2860(02)00617-8.

[52] Zheng, K., L. Zhu, Y.-T. Li, Z.-Y. Wu, C.-W. Yan, Synthesis and crystal structure of new dicopper(II) complexes having asymmetric N,N'-bis(substituted)oxamides with DNA/protein binding ability: In vitro anticancer activity and molecular docking studies, *Journal of Photochemistry and Photobiology B: Biology.* 149 (2015) 129–142. https://doi.org/10.1016/j.jphotobiol.2015.05.014.

[53] Hall, J., D. Hüsken, R. Häner, Sequence-Specific Cleavage of RNA Using Macrocyclic Lanthanide Complexes Conjugated to Oligonucleotides: A Structure Activity Study, *Nucleosides and Nucleotides.* 16 (1997) 1357–1368. https://doi.org/10.1080/07328319708006187.

[54] Liu, X., Y. Li, Y. Ci, Interaction of the La^{3+}-Morin Complex with DNA, *Analytical Sciences.* 13 (1997) 939–944. https://doi.org/10.2116/analsci.13.939.

[55] Branum, M.E., L. Que Jr., Double-strand DNA hydrolysis by dilanthanide complexes, *JBIC.* 4 (1999) 593–600. https://doi.org/10.1007/s007750050382.

[56] Zhou, J., L. Wang, J. Wang, N. Tang, Synthesis, characterization, antioxidative and antitumor activities of solid quercetin rare earth(III) complexes, *Journal of Inorganic Biochemistry.* 83 (2001) 41–48. https://doi.org/10.1016/S0162-0134(00)00128-8.

[57] Zhou, J.-H., S.-Q. Xia, J.-R. Chen, X.-S. Wang, B.-W. Zhang, H.-J. Zhang, P. Zou, X.-C. Ai, J.-P. Zhang, Surface binding and improved photodamage of the lanthanum ion complex of hypocrellin A to calf thymus DNA, *Journal of Photochemistry and Photobiology A: Chemistry.* 165 (2004) 143–147. https://doi.org/10.1016/j.jphotochem.2004.03.015.

[58] Rittich, B., A. Španová, M. Falk, M.J. Beneš, M. Hrubý, Cleavage of double stranded plasmid DNA by lanthanide complexes, *Journal of Chromatography B.* 800 (2004) 169–173. https://doi.org/10.1016/j.jchromb.2003.09.011.

[59] Qin, D., G. Qi, Z. Yang, J. Wu, Y. Liu, Fluorescence and Biological Evaluation of the La(III) and Eu(III) Complexes with 7-methoxychromone-3-carbaldehyde Benzoyl Hydrazone Schiff Base, *J Fluoresc.* 19 (2009) 409–418. https://doi.org/10.1007/s10895-008-0427-x.

[60] Ansari, A.A., R.K. Sharma, Synthesis and Characterization of a Biologically Active Lanthanum(III)–Catechin Complex and DNA Binding Spectroscopic Studies, *Spectroscopy Letters.* 42 (2009) 178–185. https://doi.org/10.1080/00387010902827718.

[61] Mohanan, K., R. Aswathy, L.P. Nitha, N.E. Mathews, B.S. Kumari, Synthesis, spectroscopic characterization, DNA cleavage and antibacterial studies of a novel

tridentate Schiff base and some lanthanide(III) complexes, *Journal of Rare Earths.* 32 (2014) 379–388. https://doi.org/10.1016/S1002-0721(14)60081-8.

[62] Asadi, Z., H. Mosallaei, M. Sedaghat, R. Yousefi, Competitive binding affinity of two lanthanum(III) macrocycle complexes toward DNA and bovine serum albumin in water, *J Iran Chem Soc.* 14 (2017) 2367–2385. https://doi.org/10.1007/s13738-017-1172-3.

[63] Asadi, Z., N. Nasrollahi, H. Karbalaei-Heidari, V. Eigner, M. Dusek, N. Mobaraki, R. Pournejati, Investigation of the complex structure, comparative DNA-binding and DNA cleavage of two water-soluble mono-nuclear lanthanum(III) complexes and cytotoxic activity of chitosan-coated magnetic nanoparticles as drug delivery for the complexes, Spectrochimica Acta Part A: *Molecular and Biomolecular Spectroscopy.* 178 (2017) 125–135. https://doi.org/10.1016/j.saa.2017.01.037.

[64] Kostova, I., I. Manolov, S. Konstantinov, M. Karaivanova, Synthesis, physicochemical characterisation and cytotoxic screening of new complexes of cerium, lanthanum and neodymium with Warfarin and Coumachlor sodium salts, *European Journal of Medicinal Chemistry.* 34 (1999) 63–68. https://doi.org/10.1016/S0223-5234(99)80041-5.

[65] Manolov, I., I. Kostova, S. Konstantinov, M. Karaivanova, Synthesis, physicochemical characterization and cytotoxic screening of new complexes of cerium, lanthanum and neodymium with Niffcoumar sodium salt, *European Journal of Medicinal Chemistry.* 34 (1999) 853–858. https://doi.org/10.1016/S0223-5234(99)00207-X.

[66] Kostova, I., R. Kostova, G. Momekov, N. Trendafilova, M. Karaivanova, Antineoplastic activity of new lanthanide (cerium, lanthanum and neodymium) complex compounds, *Journal of Trace Elements in Medicine and Biology.* 18 (2005) 219–226. https://doi.org/10.1016/j.jtemb.2005.01.002.

[67] Kostova, I., G. Momekov, T. Tzanova, M. Karaivanova, Synthesis, Characterization, and Cytotoxic Activity of New Lanthanum(III) Complexes of Bis-Coumarins, *Bioinorganic Chemistry and Applications.* 2006 (2006) e25651. https://doi.org/10.1155/BCA/2006/25651.

[68] Kostova, I., V.K. Rastogi, W. Kiefer, A. Kostovski, New Lanthanum (III) Complex – Synthesis, Characterization, and Cytotoxic Activity, *Archiv Der Pharmazie.* 339 (2006) 598–607. https://doi.org/10.1002/ardp.200600077.

[69] Kostova, I., G. Momekov, Synthesis, characterization and cytotoxicity evaluation of new cerium(III), lanthanum(III) and neodymium(III) complexes, *Applied Organometallic Chemistry.* 21 (2007) 226–233. https://doi.org/10.1002/aoc.1205.

[70] Kostova, I., N. Trendafilova, G. Momekov, Theoretical, spectral characterization and antineoplastic activity of new lanthanide complexes, *Journal of Trace Elements in Medicine and Biology.* 22 (2008) 100–111. https://doi.org/10.1016/j.jtemb.2007.10.005.

[71] Yang, Z.-Y., B.-D. Wang, Y.-H. Li, Study on DNA-binding properties and cytotoxicity in L1210 of La(III) complex with PMBP-isonicotinoyl hydrazone, *Journal of Organometallic Chemistry.* 691 (2006) 4159–4166. https://doi.org/10.1016/j.jorganchem.2006.06.002.

[72] Hussain, A., S. Gadadhar, T.K. Goswami, A.A. Karande, A.R. Chakravarty, Photoactivated DNA cleavage and anticancer activity of pyrenyl-terpyridine lanthanide complexes, *European Journal of Medicinal Chemistry.* 50 (2012) 319–331. https://doi.org/10.1016/j.ejmech.2012.02.011.

[73] Wang, Z.-M., H.-K. Lin, S.-R. Zhu, T.-F. Liu, Y.-T. Chen, Spectroscopy, cytotoxicity and DNA-binding of the lanthanum(III) complex of an l-valine derivative of 1,10-phenanthroline, *Journal of Inorganic Biochemistry.* 89 (2002) 97–106. https://doi.org/10.1016/S0162-0134(01)00395-6.

[74] Biersack, B., R. Diestel, C. Jagusch, F. Sasse, R. Schobert, Metal complexes of natural melophlins and their cytotoxic and antibiotic activities, *Journal of Inorganic Biochemistry.* 103 (2009) 72–76. https://doi.org/10.1016/j.jinorgbio.2008.09.005.

[75] Durgo, K., I. Halec, I. Šola, J. Franekić, Genotoxic effect of quercetin and lanthanum complex on human cervical cancer cells, *Archives for Occupational Hygiene and Toxicology.* 62 (2011) 221–226. https://doi.org/10.2478/10004-1254-62-2011-2122.

[76] Heffeter, P., M.A. Jakupec, W. Körner, S. Wild, N.G. von Keyserlingk, L. Elbling, H. Zorbas, A. Korynevska, S. Knasmüller, H. Sutterlüty, M. Micksche, B.K. Keppler, W. Berger, Anticancer activity of the lanthanum compound [tris(1,10-phenanthroline)lanthanum(III)]trithiocyanate (KP772; FFC24), *Biochemical Pharmacology.* 71 (2006) 426–440. https://doi.org/10.1016/j.bcp.2005.11.009.

[77] Shahraki, S., F. Shiri, M. Heidari Majd, S. Dahmardeh, Anti-cancer study and whey protein complexation of new lanthanum(III) complex with the aim of achieving bioactive anticancer metal-based drugs, *Journal of Biomolecular Structure and Dynamics.* 37 (2019) 2072–2085. https://doi.org/10.1080/07391102.2018.1476266.

[78] Abo El-Maali, N., A.Y. Wahman, A.A.M. Aly, A.Y. Nassar, D.M. Sayed, Anticancer activity of lanthanum (III) and europium (III) 5-fluorouracil complexes on Caco-2 cell line, *Applied Organometallic Chemistry.* 34 (2020) e5594. https://doi.org/10.1002/aoc.5594.

[79] Wang, B., Z.-Y. Yang, P. Crewdson, D. Wang, Synthesis, crystal structure and DNA-binding studies of the Ln(III) complex with 6-hydroxychromone-3-carbaldehyde benzoyl hydrazone, *Journal of Inorganic Biochemistry.* 101 (2007) 1492–1504. https://doi.org/10.1016/j.jinorgbio.2007.04.007.

[80] Zhou, J., L. Wang, J. Wang, N. Tang, Synthesis, characterization, antioxidative and antitumor activities of solid quercetin rare earth(III) complexes, *Journal of Inorganic Biochemistry.* 83 (2001) 41–48. https://doi.org/10.1016/S0162-0134(00)00128-8.

[81] Hassan, S.S., E.F. Mohamed, Antimicrobial, antioxidant and antitumor activities of Nano-Structure Eu (III) and La (III) complexes with nitrogen donor tridentate ligands, *Applied Organometallic Chemistry.* 34 (2020) e5258. https://doi.org/10.1002/aoc.5258.

[82] Gonca, S., S. Özdemir, A. Tekgul, C. Ünlü, K. Ocakoğlu, N. Dizge, Synthesis and characterization of perovskite type of La1-xBaxMnO3 nanoparticles with

investigation of biological activity, *Advanced Powder Technology*. 33 (2022). https://doi.org/10.1016/j.apt.2021.10.038.

[83] Avaji, P.G., B.N. Reddy, S.A. Patil, P.S. Badami, Synthesis, spectral characterization, biological and fluorescence studies of lanthanum(III) complexes with 3-substituted-4-amino-5-hydrazino-1,2,4-triazole Schiff bases, *Transition Met Chem*. 31 (2006) 842–848. https://doi.org/10.1007/s11243-006-0066-5.

[84] Avaji, P.G., S.A. Patil, P.S. Badami, Synthesis, spectral, thermal, solid state d.c. electrical conductivity and biological studies of lanthanum(III) and thorium(IV) complexes with thiocarbohydrazone, *Transition Met Chem*. 32 (2007) 379–386. https://doi.org/10.1007/s11243-006-0178-y.

[85] El-Gamel, N.E.A. Thermal studies, structural characterization, and antimicrobial evaluation of coordinated metal complexes containing salen moiety, *Monatsh Chem*. 144 (2013) 1627–1634. https://doi.org/10.1007/s00706-013-1049-9.

[86] Andotra, S., N. Kalgotra, S.K. Pandey, Syntheses, Characterization, Thermal, and Antimicrobial Studies of Lanthanum(III) Tolyl/Benzyldithiocarbonates, *Bioinorganic Chemistry and Applications*. 2014 (2014) e780631. https://doi.org/10.1155/2014/780631.

[87] Köse, M., G. Ceyhan, E. Atcı, V. McKee, M. Tümer, Synthesis, structural characterization and photoluminescence properties of a novel La(III) complex, *Journal of Molecular Structure*. 1088 (2015) 129–137. https://doi.org/10.1016/j.molstruc.2015.02.023.

[88] Alghool, S., H.F. Abd El-Halim, M.S. Abd El-sadek, I.S. Yahia, L.A. Wahab, Synthesis, thermal characterization, and antimicrobial activity of lanthanum, cerium, and thorium complexes of amino acid Schiff base ligand, *J Therm Anal Calorim*. 112 (2013) 671–681. https://doi.org/10.1007/s10973-012-2628-4.

[89] Niroomand, S., M. Khorasani-Motlagh, M. Noroozifar, S. Jahani, A. Moodi, Photochemical and DFT studies on DNA-binding ability and antibacterial activity of lanthanum(III)-phenanthroline complex, *Journal of Molecular Structure*. 1130 (2017) 940–950. https://doi.org/10.1016/j.molstruc.2016.10.076.

[90] Patil, S.S., B.P. Langi, M.N. Gurav, D.K. Patil, Synthesis, Physical and Spectral Investigations and Biological Studies of Mixed Ligand Lanthanum Complexes, *Annals of the Romanian Society for Cell Biology*. (2021) 10559–10569.

[91] Köse, M., G. Ceyhan, E. Atcı, V. McKee, M. Tümer, Synthesis, structural characterization and photoluminescence properties of a novel La(III) complex, *Journal of Molecular Structure*. 1088 (2015) 129–137. https://doi.org/10.1016/j.molstruc.2015.02.023.

[92] Horrocks Jr., W.D., Schmidt, D.R. Sudnick, C. Kittrell, R.A. Bernheim, Laser-induced lanthanide ion luminescence lifetime measurements by direct excitation of metal ion levels. A new class of structural probe for calcium-binding proteins and nucleic acids, *ACS Publications*. (2002). https://doi.org/10.1021/ja00449a079.

[93] Horrocks Jr., W.D., Sudnick D.R. Lanthanide ion probes of structure in biology. Laser-induced luminescence decay constants provide a direct measure of the number of metal-coordinated water molecules, *ACS Publications*. (2002). https://doi.org/10.1021/ja00496a010.

[94] Bkouche-Waksman, I., J. Guilhem, C. Pascard, B. Alpha, R. Deschenaux, J.-M. Lehn, Crystal Structures of the Lanthanum(III), Europium(III), and Terbium(III) Cryptates of Tris(bipyridine) Macrobicyclic Ligands, *Helvetica Chimica Acta.* 74 (1991) 1163–1170. https://doi.org/10.1002/hlca.19910740603.

[95] Zhu, Y.-J., J.-X. Chen, W.-H. Zhang, Z.-G. Ren, Y. Zhang, J.-P. Lang, S.-W. Ng, Syntheses, crystal structures and luminescent properties of two novel lanthanide/4-pya complexes: [Ln(4-pya)3(H2O)2]2 (Ln = Eu, La; 4-pya=trans-4-pyridylacrylate), *Journal of Organometallic Chemistry.* 690 (2005) 3479–3487. https://doi.org/10.1016/j.jorganchem.2005.04.041.

[96] Xiong, D.-B., M.-R. Li, W. Liu, H.-H. Chen, X.-X. Yang, J.-T. Zhao, Synthesis, structure and luminescence property of two lanthanum phosphite hydrates: La2(H2O)x(HPO3)3 (x=1,2), *Journal of Solid State Chemistry.* 179 (2006) 2571–2577. https://doi.org/10.1016/j.jssc.2006.04.035.

[97] Tang, K.-Z., J. Zhang, Y. Tang, W.-S. Liu, M.-Y. Tan, Y.-X. Sun, Preparation, crystal structure and luminescent properties of the 3-D netlike supramolecular lanthanide picrate complexes with 2,2'-[1,2-phenylenebis(oxy)]bis(N-benzylacetamide), *Inorganica Chimica Acta.* 359 (2006) 1207–1214. https://doi.org/10.1016/j.ica.2005.10.004.

[98] Dou, W., J.-N. Yao, W.-S. Liu, Y.-W. Wang, J.-R. Zheng, D.-Q. Wang, Crystal structure and luminescent properties of new rare earth complexes with a flexible Salen-type ligand, *Inorganic Chemistry Communications.* 10 (2007) 105–108. https://doi.org/10.1016/j.inoche.2006.09.017.

[99] Galyametdinov, Y.G., A.A. Knyazev, V.I. Dzhabarov, T. Cardinaels, K. Driesen, C. Görller-Walrand, K. Binnemans, Polarized Luminescence from Aligned Samples of Nematogenic Lanthanide Complexes, *Advanced Materials.* 20 (2008) 252–257. https://doi.org/10.1002/adma.200701714.

[100] Regulacio, M.D., M.H. Pablico, J.A. Vasquez, P.N. Myers, S. Gentry, M. Prushan, S.-W. Tam-Chang, S.L. Stoll, Luminescence of Ln(III) Dithiocarbamate Complexes (Ln = La, Pr, Sm, Eu, Gd, Tb, Dy), *Inorg. Chem.* 47 (2008) 1512–1523. https://doi.org/10.1021/ic701974q.

[101] Li, Y., Y. Zhao, Intramolecular Energy Transfer and Co-luminescence Effect in Rare Earth Ions (La, Y, Gd and Tb) Doped with Eu3+ β-diketone Complexes, *J Fluoresc.* 19 (2009) 641–647. https://doi.org/10.1007/s10895-008-0456-5.

[102] Roof, I.P., T.-C. Jagau, W.G. Zeier, M.D. Smith, H.-C. zur Loye, Crystal Growth of a New Series of Complex Niobates, LnKNaNbO5 (Ln = La, Pr, Nd, Sm, Eu, Gd, and Tb): Structural Properties and Photoluminescence, *Chem. Mater.* 21 (2009) 1955–1961. https://doi.org/10.1021/cm9003245.

[103] Granadeiro, C.M., R.A.S. Ferreira, P.C.R. Soares-Santos, L.D. Carlos, H.I.S. Nogueira, Lanthanopolyoxometalates as Building Blocks for Multiwavelength Photoluminescent Organic–Inorganic Hybrid Materials, *European Journal of Inorganic Chemistry.* 2009 (2009) 5088–5095. https://doi.org/10.1002/ejic.200900615.

[104] Hu, M., H.-F. Li, J.-Y. Yao, Y. Gao, Z.-L. Liu, H.-Q. Su, Hydrothermal synthesis and characterization of two 2-D lanthanide-2,2'-bipyridine-3,3'-dicarboxylate

coordination polymers based on zigzag chains, *Inorganica Chimica Acta.* 363 (2010) 368–374. https://doi.org/10.1016/j.ica.2009.11.001.

[105] Chai, X.-C., Y.-Q. Sun, R. Lei, Y.-P. Chen, S. Zhang, Y.-N. Cao, H.-H. Zhang, A Series of Lanthanide Frameworks with a Flexible Ligand, N,N'-Diacetic Acid Imidazolium, in Different Coordination Modes, *Crystal Growth & Design.* 10 (2010) 658–668. https://doi.org/10.1021/cg901075r.

[106] Yao, M., Y. Li, M. Hossu, A.G. Joly, Z. Liu, Z. Liu, W. Chen, Luminescence of Lanthanide–Dimethyl Sulfoxide Compound Solutions, *J. Phys. Chem. B.* 115 (2011) 9352–9359. https://doi.org/10.1021/jp202350p.

[107] Wang, W., Y. Huang, T. Li, G.-N. Lu, Y. Yang, N. Tang, H.-B. Song, Dinuclear rare-earth picrate complexes based on a multidentate amide ligand: syntheses, crystal structures, and luminescence properties, *Journal of Coordination Chemistry.* 65 (2012) 4041–4053. https://doi.org/10.1080/00958972.2012.731503.

[108] Bennett, S.D., B.A. Core, M.P. Blake, S.J.A. Pope, P. Mountford, B.D. Ward, Chiral lanthanide complexes: coordination chemistry, spectroscopy, and catalysis, *Dalton Trans.* 43 (2014) 5871–5885. https://doi.org/10.1039/C4DT00114A.

[109] Yin, X., J.-N. Tang, G.-H. Pan, W.-M. Tian, W.-J. Xu, Z.-J. Huang, Hydrothermal Synthesis, Structure, Properties of a Novel Supramolecular Complex: [LaCd(bpy)4(NO3)5], *Molecular Crystals and Liquid Crystals.* 606 (2015) 208–215. https://doi.org/10.1080/15421406.2014.905318.

[110] Nedilko, S.G., O. Chukova, V. Chornii, V. Degoda, K. Bychkov, K. Terebilenko, M. Slobodyanik, Luminescence properties of the new complex La,BiVO4:Mo,Eu compounds as materials for down-shifting of VUV–UV radiation, *Radiation Measurements.* 90 (2016) 282–286. https://doi.org/10.1016/j.radmeas.2016.02.027.

[111] Allendorf, M.D., C.A. Bauer, R.K. Bhakta, R.J.T. Houk, Luminescent metal–organic frameworks, *Chem. Soc. Rev.* 38 (2009) 1330–1352. https://doi.org/10.1039/B802352M.

[112] Amoroso, A.J., S.J.A. Pope, Using lanthanide ions in molecular bioimaging, *Chem. Soc. Rev.* 44 (2015) 4723–4742. https://doi.org/10.1039/C4CS00293H.

[113] Wang, W., Y. Huang, T. Li, G.-N. Lu, Y. Yang, N. Tang, H.-B. Song, Dinuclear rare-earth picrate complexes based on a multidentate amide ligand: syntheses, crystal structures, and luminescence properties, *Journal of Coordination Chemistry.* 65 (2012) 4041–4053. https://doi.org/10.1080/00958972.2012.731503.

Biographical Sketches

Vibha Vinayakumar Bhat, PhD

Affiliation: Department of Chemistry, M S Ramaiah College of Arts, Science and Commerce, MSR Nagar, Bengaluru, Karnataka, India

Education: MSc, PhD

Business Address: M S Ramaiah College of Arts, Science and Commerce, MSR Nagar, Bengaluru, Karnataka, India

Research and Professional Experience: 7 years Research and 4 years Professional Experience

Professional Appointments: Assistant Professor

Honors:
- Awardee of Prof. B. Sanjeeva Rao Memorial Gold Medal award
- Awardee of Prof. G. K. Narayana Reddy Gold Medal award
- Awardee of Prof. H. Sankegowda Cash Prize

Publications from the Last 3 Years:
1. Vinayakumar, Bhat Vibha, P. R. Chetana, Fluorescence studies of Lanthanum (III) complexes of N, N' bis-(alkyl/aryl)-substituted oxamides and phenanthroline bases, *Res. J. Chem. Env.,* 2020, 24, 88 – 95.
2. Chetana, P. R., D. R. Navya, Vibha Vinayakumar Bhat, B. S. Srinatha, Mohan A. Dhale, Studies on DNA interactions and Biological Activities of Lanthanum(III) Complexes with 4-quinoline terpyridine and 1,10-phenanthroline. *Asian Journal of Chemistry,* 2019, 6, 1265-1274.
3. Chetana, P. R., Vibha Vinayakumar Bhat, Mohan A. Dhale, Heterobinuclear complexes of lanthanum(III) using bridging N,N'-bis(2-pyridylmethyl)oxamide and terminal 1,10-phenanthroline: Syntheses, characterization and DNA interactions. *Int. J. Pharm. Sci. Drug Res.* 2018; 10(6): 460-473.
4. Chetana, P. R., Vibha Vinayakumar Bhat, Mohan A. Dhale, DNA interactions, antibacterial and antioxidant studies of newly synthesized

lanthanum (III) complexes using N,N′-bis(3-pyridylmethyl) oxamide and N,N-heterocyclic bases. *Int. J. Pharm. Sci. Rev. Res.* 2018, 49, 86-99.
5. Bhat, Vibha Vinayakumar, P. R. Chetana, Mohan A Dhale and Sreejit Soman, Synthesis, characterization, in vitro biological evaluation and molecular docking studies of newly synthesized mononuclear lanthanum(III) complexes of N,N'-bis(2- aminoethyl)oxamide and phenanthroline bases, *Journal of Molecular Structure* (Under Revision).

P. R. Chetana, PhD

Affiliation: Department of Chemistry, Central College Campus, Bengaluru City University, Bengaluru, India

Education: MSc, PhD

Business Address: Department of Chemistry, Central College Campus, Bengaluru City University, Bengaluru, India

Research and Professional Experience: 20 years of Research and 25 years of Professional Experience

Professional Appointments: Professor

Honors: Awardee of Prof. G. K. Narayana Reddy Gold Medal award

Publications from the Last 3 Years:
1. Somashekar, M. N., P. R. Chetana, B. S.Chethan, H. R. Rajegowda, M. A. Cooper, Z. M. Ziora, N. K. Lokanath, P. S. Sujan Ganapathy, B. S. Srinatha, Synthesis and characterization of Zinc(II) complex with ONO donor type new phenylpropanehydrazide based ligand: Crystal structure, Hirshfeld surface analysis, DFT, energy frameworks and molecular docking, *Journal of Molecular Structure,* 2022, 1255, 132429.
2. Vinayakumar, Bhat Vibha, P. R. Chetana, Fluorescence studies of Lanthanum (III) complexes of N, N′ bis-(alkyl/aryl)-substituted oxamides and phenanthroline bases, *Res. J. Chem. Env.,* 2020, 24, 88 – 95.
3. Chetana, P. R., D. R. Navya, Vibha Vinayakumar Bhat, B. S. Srinatha, Mohan A. Dhale, Studies on DNA interactions and Biological Activities

of Lanthanum(III) Complexes with 4-quinoline terpyridine and 1,10-phenanthroline. *Asian Journal of Chemistry,* 2019, 6, 1265-1274.
4. Chetana, P. R., Vibha Vinayakumar Bhat, Mohan A. Dhale, Heterobinuclear complexes of lanthanum(III) using bridging N,N'-bis(2-pyridylmethyl)oxamide and terminal 1,10-phenanthroline: Syntheses, characterization and DNA interactions. *Int. J. Pharm. Sci. Drug Res.* 2018; 10(6): 460-473
5. Chetana, P. R., Vibha Vinayakumar Bhat, Mohan A. Dhale, DNA interactions, antibacterial and antioxidant studies of newly synthesized lanthanum (III) complexes using N,N'-bis(3-pyridylmethyl) oxamide and N,N-heterocyclic bases. *Int. J. Pharm. Sci. Rev. Res.* 2018, 49, 86-99.
6. Bhat, Vibha Vinayakumar, P. R. Chetana, Mohan A Dhale and Sreejit Soman, Synthesis, characterization, in vitro biological evaluation and molecular docking studies of newly synthesized mononuclear lanthanum(III) complexes of N,N'-bis(2- aminoethyl)oxamide and phenanthroline bases, *Journal of Molecular Structure (Under Revision).*

Chapter 2

The Effects of Lanthanum on Aquatic Organisms

C. Figueiredo[1,*], P. Brito[1,2], M. Caetano[1,2] and J. Raimundo[1,2]

[1]CIIMAR – Interdisciplinary Centre of Marine and Environmental Research, Matosinhos, Portugal
[2]Division of Oceanography and Marine Environment, IPMA – Portuguese Institute for Sea and Atmosphere, Algés, Portugal

Abstract

Our high-technology society is increasingly reliant on technology-critical elements (TCE), such as the Rare Earth Elements (REE). These elements are vital to "environmental-friendly" technologies, agriculture, and medical applications (e.g., MRI contrasts). With growing REE applications, emission to aquatic ecosystems rises through mining, ore processing, fossil fuel burn, and disposal of end life products (e.g., e-waste). Remarkably, maximum REE levels and discharges to the aquatic environment present no regulatory thresholds as evidence on the risks of increasing REE availability is limited. Nevertheless, REE are contaminants of emerging environmental concern that enter the aquatic medium through manifold ways. Lanthanum (La) is one of the most abundant and reactive REE in the environment and, thus, is relevant to understand its fate, bioavailability, bioaccumulation, and toxicological limits. This chapter analyses previously available scientific information on the La bioaccumulation, elimination, and ecotoxicology, focusing on freshwater and marine organisms. Lanthanum impacts biological responses and traits in multiple ways and may show interactive effects

---

[*] Corresponding Author's Email: cafigueiredo@fc.ul.pt.

In: What to Know about Lanthanum
Editor: Catherine C. Bradley
ISBN: 979-8-88697-615-1
© 2023 Nova Science Publishers, Inc.

on marine biota coupled with other chemical and abiotic stressors. We will shortly explain the organisms' bioaccumulation and elimination capacity, while providing an overview of their hampering effects, at different levels of biological organization from microorganisms to plants, invertebrates, and vertebrates. Lanthanum can be detected in almost all biota although information on its speciation is regularly not available. Publications on trophic transfer of La was not found and hence biomagnification evidence is lacking. Nevertheless, La ecotoxicological outcomes are dispersed, mainly due to species specific differences and exposure conditions disparity that include exposure to distinct La compounds. This chapter highlights an urgent need for further studies on this emerging problem.

Keywords: rare earth elements, Lanthanum, bioaccumulation, elimination, ecotoxicity, review

1. Introduction

The peculiar Rare Earth Elements (REE) are a coherent group of seventeen strongly connected elements, encompassing the Lanthanide Group (Ln) and Scandium (Sc) and Yttrium (Y). These chemical group was constituted due to their incredibly analogous chemical behavior. The remarkable physical and chemical similarity occurs since the Lanthanides (Ln) are generally trivalent ($^{3+}$), and the radii of Ln^{3+} decreases with increasing atomic number, in the identified lanthanide contraction phenomena (Hu et al., 2017). The name REE has historically misled scientists to think they are rare, when natural REE concentration in the Earth's crust may vary between 150 and 200 mg kg^{-1}, being hence more common than the well-studied elements copper (Cu), lead (Pb), gold (Au), and platinum (Pt) (Humphries, 2010). Lanthanum (La) is the first element of the Ln series and is one of the most abundant REE in the environment (Herrmann et al., 2016).

Curiously, their characteristics marked them much more technological, environmental, and economically essential than their anonymity might signal. With expanding modern technological applications in the past decades, their consumption as rose to an unprecedented rate, growing their demand on a worldwide scale. They are applied from rechargeable batteries, super magnets, computer memories and screens, mobile phones, LED lighting, fluorescent materials, solar panels, to magnetic resonance imaging (MRI) agents (Balaram, 2019). Moreover, several REE, including La, are of the most crucial importance to produce "green technologies," namely electric and hybrid vehicles batteries, wind turbine generators, low-energy lighting, fuel cells and magnetic refrigeration (Atwood, 2013). With usage of "green technologies" on the rise, the exponential world demand of these emerging elements is expected to be upheld in the coming years.

2. Lanthanum in Aquatic Environments

The REE can be detected in virtually all biota and the uptake pattern commonly follows the Oddo–Harkins rule with La usually being the second most common, after cerium (Ce) (Weltje et al., 2002). The leading REE continental sources are volcanic emissions and parent rock eroding, as of rain and watercourses. Whilst hydrothermal crusts and vents, and Fe-Mn nodules, constitute the major natural sources in the deep-sea. Their extraction, which includes mining and purification activities, are the major anthropogenic sources. These processes may provoke soil erosion, land use shift, flooding, and water and air pollution which ultimately leads to alarming biodiversity loss (Carpenter et al., 2015). For example, acid mine drainage is known to lead to discharges of low pH and La high concentrated effluents (Burch et al., 2011). Hence, the aquatic environment forms an effective sink of the prospective contaminants REE. In fact, previous studies have shown anthropogenic introduced La levels in the hydrosphere, by industrial emissions (Kulaksız and Bau, 2011). Furthermore, urban wastewater treatment systems, and domestic and industrial effluents impact these elements abundance. These anthropogenic elements are discharged in the aquatic environment in a superior soluble and reactive form, then the naturally occurring elements, which may also increase their bioavailability. The impact of REEs on aquatic organisms is of particular concern due to the importance of these ecosystems, however, to date, these have been majorly neglected as highlighted by the little amount of research articles on these topics.

Enhanced demand and briefer lifespans of various electronic products have led to disturbing creation of electronic waste (e-waste). While roughly 40 million metric tons of e-waste is generated globally, which constitutes about 5% of the worldwide waste (Hazra et al., 2019), the recycling of the elements needed for its manufacture, is practically not employed (Binnemans et al., 2021). Besides unoptimized techniques, this recycling processes apply extremely aggressive solvents at very high temperatures and pressures, with great impacts to the environment. Furthermore, the dismantling, open storage and burning of e-waste is known to release its chemical components into the environment (Uchida et al., 2018). With REE utilization in other areas such as agriculture, forestry, animal husbandry and aquaculture, used as bactericides or fertilizers (Gwenzi et al., 2018), the exponential transfer to aquatic ecosystems is expected to be maintained in coming years. The ever-growing REE market has triggered a great ecotoxicological concern, yet further studies are still needed for the establishment of toxicity and safety parameters. The high levels of contamination may impose damaging effects on biodiversity, ecosystem functioning and ultimately human health.

While the number of studies on REE, including La bioaccumulation and toxic effects, have increased in recent years (Figure 1), due to the growing importance of this emerging element, few studies exist, particularly on vertebrate species (Figure 2) with quite puzzling information and in some cases describing conflicting results. This highlights the usefulness of this chapters, that bring together the distinct studies throughout time.

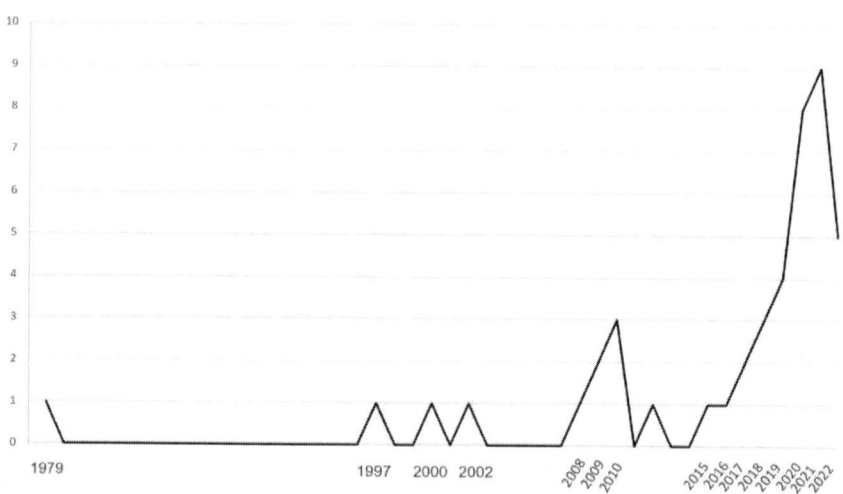

Figure 1. Number of La exposure on aquatic organisms' research papers published per year.

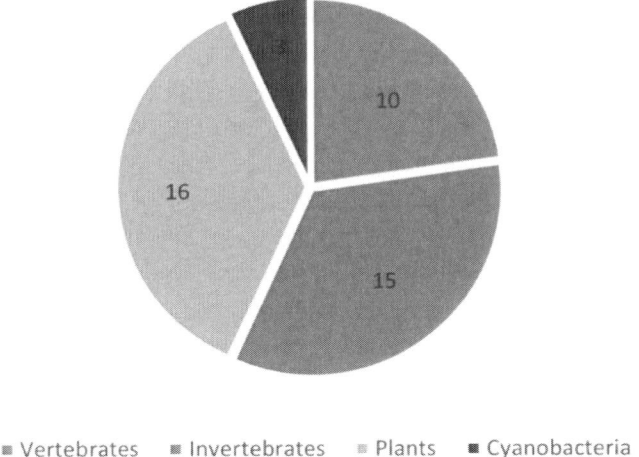

Figure 2. Summary of number of model organisms in La uptake, elimination and/or ecotoxicity research papers.

3. Overview on Lanthanum Bioaccumulation and Ecotoxicity on Aquatic Organisms

A vast search was prosecuted towards the collection of research articles of laboratory La exposure. A comprehensive search of peer-reviewed papers occurred in April 2022, at the search engines Web of Science (www.webofscience.com); Google Scholar (www.scholar.google.com) and ScienceDirect (www.sciencedirect.com).

An overview of La exposure experiments details is shown in Table 1.

Metal speciation, the physicochemical water parameters and the organisms' physiological traits are examples of the various abiotic and biotic factors that can alter the bioavailability of metals. The bioavailability of La, and the remaining REE, is strongly related to its speciation, which in turn depends on a wide assortment of features such as pH, salinity, major ions, and organic matter (Liang et al., 2005). Furthermore, salinity alters REE speciation which may impact these elements uptake (Figueiredo et al., 2022b). Globally, REE are present as free ions, carbonates, and hydroxyl complexes, depending on pH. Furthermore, the REE can form complexes with cations, anions, neutral ions, and organic and inorganic ligands. With increasing atomic number, the stability constants formations of REE as carbonates also increases. Furthermore, the complexation with phosphate, sulfate and hydro-oxide is simpler for HREE, than LREE as La. Although REE speciation in water samples has been modeled before, this has been done majorly considering inorganic matter and speciation with organic matter has been seriously disregarded. This assessment constitutes a great challenge as the interpretation of REE complexes is considerably intricate since it depends on geographical traits, industrialization, weathering, and element transport processes, which has led to great inconsistency on La uptake patterns described in the literature. Moreover, the consumption of seafood in several countries is ecologically and economically key, therefore, the understanding of bioaccumulation capabilities and induced ecotoxicological effects is of the uttermost importance as this human consumption may lead to unknown health impairments. In fact, previous studies have described health hazards as a consequence of regular exposure to REE (e.g., Rim, 2016).

Recent reviews on published data on the effects of REE state that most scientific articles focus on the impact of REE on aquatic organisms and environments, while La is the most studied REE and its ecotoxicological effects are one of the most described in literature, followed by Cerium (Ce) and Gadolinium (Gd) (Gonzalez et al., 2014; Herrmann et al., 2016; Malhotra et al., 2020). Therefore, this chapter will bring together previously publish data on aquatic *organisms exposed to La, its bioaccumulation and elimination capacity, while providing an overview of its hampering effects, at different levels of biological organization.*

The bioaccessibility and bioavailability of a given element are key for the knowledge on the total metal that is accumulated. Bioaccessible elements are theoretically available, however may be out of reach to the organism, while the bioavailable portion is the fraction able to enter the organism. The La accumulation can occur from body surface contact and/or prey consumption and toxicity arise when the detoxification and elimination rates are surpassed by the rate of uptake (Luoma and Rainbow, 2005). Overall, information found in the literature states that La may be accumulated, in a tissue specific manner, while being dependent on exposure concentration and several toxicological outcomes have been described, such as liver impaired function and brain damage. These adverse effects may be improved in combination with other pollutants and in climate change scenarios. Although a couple of papers described La toxicity in zebrafish, which model species has been widely applied in aquatic ecotoxicity assays with a broad array of pollutants (e.g., pesticides, heavy metals, solvents, etc.), no hotspot aquatic species has been studied and proposed.

Table 1. Summary of La exposure experiments on aquatic organisms. Asterisks represent research articles on the effects of La under climate change variables (i.e., ocean warming, acidification and/or salinity variations)

		Model	Composite	Exposure Concentration	Duration	Depuration	Reference
Vertebrates	Fish	*Anguilla anguilla*	$LaCl_3$	120 ng L^{-1}	7 days	7 days	Figueiredo et al. 2018
		Anguilla anguilla	$LaCl_3$	1.5 µg L^{-1}	5 days	5 days	Figueiredo et al. 2020*
		Oncorhynchus mykiss	$LaCl_3$	0.064, 0.32, 1.6, 8 and 40 mg L^{-1}	96 h	-	Dubé et al. 2019
		Oncorhynchus mykiss	mixture of La, Ce, Nd, Sm, Pr	14, 140, 1400, 14000 µg L^{-1}	96 h	-	Hanana et al. 2021
		Carassius auratus	La ion and EDTA complex	0.015 mM	24 days	-	Chen et al. 2020
		Carassius auratus	$La(NO_3)_3$	5mM L^{-1}	5 h	-	Eddy and Bath 1979
		Gobiocypris rarus	$LaCl_3$	0.04, 0.08, 0.16, 0.32 and 0.80 mg L^{-1}	96 h	-	Hua et al. 2017
		Gobiocypris rarus	$LaCl_3$	0.04, 08, 0.16, 0.32, and 0.80 mg L^{-1}	21 days	-	Qiu et al. 2020
		Danio rerio	$La(NO_3)_3$	0, 10, 20, and 30 µmol L^{-1}	6 days	-	Liu et al. 2021
		Danio rerio	$La(NO_3)_3$	0.8, 1.6, 3.2, 6.4, 12.8 and 25.6 mg L^{-1}	48 h	-	Huang et al. 2022
Invertebrates	Bivalves	*Mytilus galloprovincialis*	$LaCl_3$	0.01, 0.1, 1, 10, 100 mg L^{-1}	96 h	-	Mestre et al. 2019
		Mytilus galloprovincialis	-	10 µg L^{-1}	28 days	-	Andrade et al. 2021*

Table 1. (Continued)

		Model	Composite	Exposure Concentration	Duration	Depuration	Reference
Invertebrates	Bivalves	*Mytilus galloprovincialis*	$LaCl_3$	0, 0.1, 1, 10 mg L^{-1}	28 days	-	Pinto et al. 2019
		Mytilus galloprovincialis	-	10 μg L^{-1}	14 days	14 days	Cunha et al. 2022
		Crassostrea gigas	La_2O_3	2.5, 5.0, 10, 20, 40 and 160 μg L^{-1}	48 h	-	Moreira et al. 2020
		Ruditapes philippinarum	$LaCl_3$	0.3 μg L^{-1} and 0.9 μg L^{-1}	6 days	-	Figueiredo et al. 2022a
		Spisula solida	$LaCl_3$	15 μg L^{-1}	7 days	7 days	Figueiredo et al. 2022c
	Echinoderms	*Paracentrotus lividus*	$LaCl_3$	from 10^{-6} to 10^{-4} M	72 h	-	Pagano et al. 2016
		Sphaerechinus granularis / *Arbacia lixula*	$LaCl_3$	from 10^{-7} to 10^{-4} M	72 h	-	Trifuoggi et al. 2017
	Crustaceans	*Daphnia carinata*	$LaCl_3$	100, 200, 400, 600, 800, and 1000 μg L^{-1}	48 h	-	Barry and Meehan 2000
		Daphnia magna	-	10, 30, 50 and 100 mg L^{-1}	72 h	-	Korkmaz et al. 2021
		Daphnia magna / *Thamnocephalus platyurus* / *Heterocypris incongruens*	$La(NO_3)_3$	6.25, 12.5, 25 and 50 mg L^{-1} acute test and 0.01, 0.1, 0.25, 0.5 and 1.0 mg L^{-1} reproduction test	72h and 21 days	-	Blinova et al. 2018

		Model	Exposure		Duration	Depuration	Reference
			Composite	Concentration			
Invertebrates	Crustaceans	*Daphnia magna*	$La(NO_3)_3$ and Phoslock ®	33, 100, 330 and 1000 µg L^{-1} and 100, 500, 5000 mg Phoslock ® L^{-1}	16 days	-	Lürling and Tolman 2010
		Daphnia magna	La_2O_3 nanoparticles	10, 50, 100, 250, 500 and 1000 mg L^{-1}	72 h	-	Balusamy et al. 2015
		Daphnia similis *Artemia salina*	La, Nd, Sm, individually and combinations	max. 300 mg L^{-1} then fractionated in seven concentrations until EC50 was determined	48 h	-	Bergsten-Torralba et al. 2020
Plants		*Chlorella vulgaris* *Raphidocelis subcapitata* *Chlorella fusca*	$La(NO_3)_3$ and Malic acid and NTA or IDA	0.010–100 µM	120 h	-	Aharchaou et al. 2020
		Chlorella sp.	La_2O_3 nanoparticles	10, 50, 100, 250, 500 and 1000 mg L^{-1}	72 h	-	Balusamy et al. 2015
		Chlorella vulgaris	$La(NO_3)_3$ and EDTR, NTA and CIT	0.5 to 4 mg mL^{-1}	50 h	-	Hao et al. 1997
		Chlorella kessleri	$La(NO_3)_3$	up to 1000 µM	48 h	-	Fujiwara et al. 2008
		Chlorella vulgaris *Phaeodactylum tricornutum*	$La(NO_3)_3$	2, 4, 6, 8, and 10 mg L^{-1}	96 h	-	Sun et al. 2019
		Skeletonema costatum	$LaCl_3$	5, 10, 20, 30, and 40 µmol L^{-1}	96 h	-	Tai et al. 2010
		Raphidocelis subcapitata	$La(NO_3)_3$	0.7-5.5 mg L^{-1}	28 days	-	Siciliano et al. 2021b
		Raphidocelis subcapitata	$LaCl_3$	0.01 and 10.22 mg L^{-1}	72 h	-	Siciliano et al. 2021a

Table 1. (Continued)

	Model	Composite	Exposure Concentration	Duration	Depuration	Reference
Plants	Desmodesmus quadricauda	$LaCl_3$ and DOTA	0.2, 0.3, 0.7, 2.3, 4.3, 13.5, 36.6, 128, 496 and 1387 nM	14 days	-	Ashraf et al. 2021
	Scenedesmus quadricauda	$La(NO_3)_3$ and EDTA	0.72, 7.2 and 72 µmol L^{-1} La and 0.269 - 26.9 µmol L^{-1} EDTA	25 days	-	Jin et al. 2009
	Fucus spiralis	La, Ce, Pr, Nd, Eu, Gd, Tb, Dy, Y	1 µmol L^{-1}	7 days	-	Costa et al. 2020
	Fucus vesiculosus				-	
	Gracilaria sp.				-	
	Osmundea pinnatifida				-	
	Ulva intestinalis				-	
	Ulva lactuca				-	
	Ulva rigida	$LaCl_3$	15 µg L-1	7 days	7 days	Figueiredo et al. 2022d*
	Lemna minor	$LaCl_3$ and EDTA	10nM La and 5.37 µM EDTA	8 days	-	Weltje et al. 2002
	Lemna minor	$La(NO_3)_3$	1, 5, and 10 mM of La and 1, 5, and 10 mM of REE (100.07 mM La, 327.57 mM Ce, 25.76 mM Pr, 0.14 mM Nd and 0.006 mM Gd)	5 days	-	Ippolito et al. 2010
	Hydrocharis dubia	$La(NO_3)_3$	0, 40, 80, 120, and 160 µM	7 days	-	Xu et al. 2012

	Model	Composite	Exposure		Duration	Depuration	Reference
			Concentration				
Cyanobacteria	*Microcystis aeruginosa*	LaCl$_3$	0.2, 2.0, 4.0, 8.0, 20, 40, and 60 μM		21 days	-	Shen et al. 2018
	Microcystis aeruginosa	Magnetite/La hydroxide (MLC-10)	0.1, 0.5 and 1 g		30 days	-	Song et al. 2021
	Microcystis aeruginosa	La(NO$_3$)$_3$	0.72, 7.2 and 72 μmol L^{-1} La and 0.269 - 26.9 μmol L^{-1} EDTA		25 days	-	Jin et al. 2009

3.1. Vertebrates

Considering the vulnerability of early life stages to a several contaminants, Figueiredo et al. (2018) exposed glass eels (*Anguilla anguilla*) to 120 ng L^{-1} of La for 7 days, along with 7 days of elimination. The La bioaccumulation was highest after 72h, and the uptake decreased even in a continuously exposed environment. The uptake was greater in the eel's viscera, followed by the skinless body and finally the head. The authors argued that this accumulation pattern could be related to a protective mechanism to deal with La neurotoxicity. In fact, enhanced AChE activity was described in the La exposed glass eels' heads. Furthermore, the authors showed a profound depression in lipid peroxidation, suggesting that La may act as a free radical scavenger. Catalase activity was also inhibited, while the quantification of GST did not show differences between control and La exposed glass eels, indicating that not all biomarkers are good proxies for La ecotoxicity. Figueiredo et al. (2020) exposed the same species and life form to 1.5 µg L^{-1} of La for five days, plus a depuration phase of the same duration, under control and a warming scenario (T=+4°C), in an attend to assess the uptake and elimination of La, and specific biochemical impairments under a climate change scenario. The results proved an enhanced La uptake and toxicity in a warming setting. Furthermore, elimination was less efficient at higher temperatures. On the contrary to the previous study, 1.5 µg L^{-1} La exposure did not affect AChE activity. Lipid peroxidation was highest in the exposure to warming and La treatment, while the head shock proteins were hampered in glass eel exposed to La, at both control temperature and warming. Hence, La impeded glass eels to proficiently prevent cellular damage, as this as more obvious in a temperature rise scenario.

Dubé et al. (2019) studied the lethal and sublethal toxicity of REEs to juvenile rainbow trout (*Oncorhynchus mykiss*). The trout was exposed to 0.064, 0.32, 1.6, 8 and 40 mg L^{-1}, of 7 REE, including La. The 96 h-LC50 of $LaPO_4$ in rainbow trout was > 63 mg L^{-1}, which brand La as a low toxic REE, in accordance with the estimated LC50 (120 mg L^{-1}). Nevertheless, La toxicity appeared to be life stage dependent as the 96-h toxicity upsurges at the fry stage reaching LC50 dramatically different levels of < 0.5 mg L^{-1} (Martin and Hickey, 2004). Also, rainbow trout eggs presented an LC50 of 0.02 mg L^{-1} in a 28-day exposure (Birge et al., 1979). Of the studied endpoints, GADD45, HSP70 and CYP1A1 were regularly impacted, which indicates that DNA damage and cell growth arrest, protein chaperones and hemoprotein-mediated xenobiotic biotransformation pathways were involved in REEs toxicity. On a different study, this same species was exposed to a mixture of the five more common REE, for 96h at concentration of 0.1, 1, 10, and 100X whereas the 1x mixture was composed of Cerium (Ce, 280 µg L^{-1}), Lanthanum (La, 140 µg L^{-1}), Neodymium (Nd, 120 µg L^{-1}), Praseodymium (Pr, 28 µg L^{-1}), and Samarium (Sm, 23 µg L^{-1}) by Hanana et al. (2021). The authors determined gene expression in the fish liver while evaluating the biomarkers in gills, due to mass limitation. The results show that genes involved in catalase, heat shock proteins 70 and glutamate dehydrogenase were upregulated, as glutathione S-transferase and metallothionein were downregulated. Finally, although three of the five studied REE, including La, presented high LC50 levels individually, their mixtures significantly impaired several genes, even at concentrations lower than the ones found naturally in the environment. Overall, having in consideration the applied exposure concentrations, the described effects are expected in juvenile fish inhabiting environments severely impacted by REE mining activities.

In a laboratory study, Chen et al. (2000) exposed goldfish *Carassius auratus* to REE with the objective of validating the use of antioxidant enzymes as biomarkers. Lanthanum was selected as a representative of light REE, and the goldfish were exposed to 0.015 mM of La and its EDTA. CAT activities of fish exposed to La-EDTA for two days decreased significantly, while the activities of fish exposed to La^{3+} were not different from controls. The authors proposed that organic complexes may enhance La bioaccumulation and toxicity. The effects of light REE La were like medium REE (Gd), however distinct from heavy REE (Y). As the previously described result was not observed for SOD, the authors state that the ecotoxicological effects depend on the species of REE and has a different impact on different enzymes. On an older article, Eddy and Bath (1979) explored the effect of La exposure on Na$^+$ and C$^-$ fluxes on the same goldfish species. The exposure to La strongly impacted the studied fluxes as La provoked enhanced influx and efflux, after 15 min both rates diminished, however effluxes remained at high levels triggering a significant net loss of ions. This loss was associated to contribute to the cause of death that arose after 10-20h La exposure (5 mM L^{-1}).

Hua et al. (2017) assessed the acute toxicity of La on the gill and liver of the rare minnow (*Gobiocypris rarus*). This freshwater fish was exposed to 0.04, 0.08, 0.16, 0.32 and 0.80 mg L^{-1} of La for a period of 21 days. The median lethal La concentration for this fish species at 96 h was established at 1.92 mg L^{-1}. The authors described that hepatological changes occurred for both studied tissues. In the gills, La exposure provoked epithelial lifting, filamentary epithelial proliferation, edema, lamellar fusion, desquamation, and necrosis. For the liver, the observed effects were dilation of sinusoids, focal congestion, pyknotic nuclei, karyohexis and karyolysis, vacuolar degeneration, and several necrosis areas. Furthermore, hypsometric analysis showed significant alteration of gill dimensions, suggesting that La elicits indirect metabolic impacts through impaired gas exchange. Qiu et al. (2020) described the La induced genotoxicity and cytotoxicity to the same freshwater fish species (*G. rarus*). Specimens were exposed during 21 days to 0.04, 08, 0.16, 0.32, and 0.80 mg L^{-1}. Exposure to La, except for 0.04 mg L^{-1}, altered the nucleus morphology. The total ratio of erythrocytes with nuclear anomalies was significantly higher in exposed fish. La was cytotoxic to erythrocytes of this fish species as significant dose-dependent alterations in erythrocyte and nucleus proportion occurred. The dimensions of erythrocyte and nuclei in exposed fish decreased.

Liu et al. (2021) exposed male zebrafish (*Danio rerio)* to 0, 10, 20, and 30 μmol L^{-1} of La, and investigated lipid deposition and Wnt10b signaling in the liver. Overall, La was accumulated in the liver and exposure to this element decreased both activities and gene expression of enzymes involved in fatty acid synthesis. The levels of total cholesterol, triglyceride, nonesterified fatty acids, and size of lipid droplets diminished with La exposure. Furthermore, La exposed male zebrafish showed impacted fatty acid composition and altered nutrient levels. Globally, the authors suggest that La accumulation alters lipid deposition in male zebrafish livers. Nevertheless, sex specific differences in relation to La uptake and toxicity are yet to be unveiled. Huang et al. (2022) also exposed *D. rerio* to 0, 0.8, 1.6, 3.2, 6.4, 12.8 and 25.6 mg L^{-1} of La, Ce and Nd. The experiment showed LC50 of La, Ce and Nd of 2.53, 2.03, and 2.76 mg L^{-1}, in that order. The scientists described several impacted biological processes, such as the metabolism of xenobiotics, oocyte meiosis, steroid biosynthesis, DNA replication, and p53 signaling pathway. These effects were similar for the three studied REE. Particularly, oocyte meiosis, progesterone mediated oocyte maturation, and steroid biosynthesis pathways, all associated with the reproductive cycle, were enhanced with exposure to all three REE, which suggests that La, Ce and Nd may disrupt the normal reproduction of this model species. This recent study contributed to unveil the possible effects of lanthanides on the reproduction of aquatic organisms, as previously described effects were limited to lethal effects, embryo damage and ecotoxicity biomarkers.

3.2. Invertebrates

3.2.1. Bivalves

Mestre et al. (2019) studied the ecotoxicity of La and Y to embryos and juveniles of *Mytilus galloprovincialis*. Embryogenesis success after 48 h, and survival of juveniles after 96 h of exposure to different concentrations of 0.01, 0.1, 1, 10, 100 mg L^{-1} of La and 0.001, 0.01, 0.1, 1, 10 mg L^{-1} of Y, independently, was accessed. Increasing concentrations of La and Y decreased the percentage of normally developed embryos as abnormal features rose. In case of juveniles, mortality rose with increasing La concentration and LC50 ranged between 10 and 100 mg L^{-1} for La. Data proved that La was more toxic than Y for this bivalve and that both elements are more toxic to developing embryos and larvae than juveniles, highlighting a 100-fold toxicity difference of La at different life stages. This study illustrates the need for further tests to be conducted on distinct lifecycle stages. Andrade et al. (2021) intended to understand the effects of salinity and La and for that the team exposed the same mussel species to 10 µg L^{-1} of La in different salinities (20, 30 and 40). The authors showed that decreased salinity triggered several biochemical impairments, as metabolism (electron transport system activity, glycogen, and total protein contents), antioxidant (superoxide dismutase, catalase and glutathione reductase), and biotransformation defenses activation (glutathione S-transferases and carboxylesterases activities), which are typically linked to hypotonic stress. This oxidative stress occurred despite La exposure. Nevertheless, only mussels exposed to La presented increased cellular damage and neurotoxicity (lipid peroxidation, protein carbonylation and acetylcholinesterase activity). Pinto et al. (2019) assessed the La toxicity on the same model, by evaluating metabolic, oxidative stress, neurotoxicity, and histopathological markers. Mussels were exposed for 28 days to increasing concentrations of La (0, 0.1, 1, 10 mg L^{-1}). La was bioaccumulated and prompted lowered metabolism and energy reserves, the antioxidant biomarkers SOD and GPx were activated, as well as GSTs. Nevertheless, oxidative damage was avoided as illustrated by lower LPO and PC levels. On another hand, inhibition of AChE demonstrated the La neurotoxicity. The calculation of histopathological indices showed the impacts of La on gonads, gills, and digestive gland. The authors suggest, having in consideration the effects of La on this edible mussel species, that its discharge should be monitored. Cunha et al. (2022) exposed the same model organisms to 10 µg L^{-1} of La and Gd, in independent trials, for 14 days, after which a same period depuration phase began. After 14 exposure days, the mussels showed higher accumulation values of La (~0.54 µg g^{-1}) than Gd (~0.15 µg g^{-1}). Exposure to these elements elicited lower energy expenditure, however after the elimination phase, the mussels were able to attain control like levels of glycogen and protein concentrations. Neurotoxicity did not occur in mussels exposed to La. This study showed how bivalves exposed to environmentally relevant concentrations of La can re-establish their biochemical performance following a recovery phase.

Regarding a different bivalve, Moreira et al. (2020) conducted a toxicity study of La and Y on the oyster *Crassostrea gigas* embryonic development. The oysters were exposed to 2.5, 5.0, 10, 20, 40 and 160 µg L^{-1} of both Y and La in separate. Here La proved to be more toxic to embryolarval development than Y with a EC50 of 6.7 to 36.1 µg L^{-1} (24 and 48 h) for La and 147 to 221.9 µg L^{-1} (24 and 48 h) for Y. The authors argued that the higher toxicity exhibited by La could have been related to higher bioavailability of its free ionic form.

In an attempt to better detail the organotrophic outline, Figueiredo et al. (2022a) exposed the Manila Clam (*Ruditapes philippinarum*) to 0.3 µg L^{-1} and 0.9 µg L^{-1} of La for 6 days and assessed the bioaccumulation pattern in different tissues: gills, digestive gland, and remaining body. This short exposure period was sampled three times, after 1, 2 and 6 days of exposure. This species accumulated La after 48h of exposure and in a dose-dependent way. When exposed to the lower concentration the enrichment pattern was gills > body > digestive gland, nevertheless, when exposed to the triple of this concentration the pattern was gills > digestive gland > body. The authors suggested that tissue portioning may be associated to exposure concentration as at a higher exposure the digestive gland surpassed the body, possibly as a detoxification mechanism.

Finally, as a study on the effects of ocean warming, acidification, and their interaction with La had never been performed, Figueiredo et al. (2022c) exposed the surf clam (*Spisula solida*) to 15 µg L^{-1} La for 7 days and performed a depuration phase of 7 further days, at control temperature and pH, warming, acidification and warming and acidification combined. Lanthanum was accumulated just after 24h at present day conditions and climate change scenarios and a 7-day elimination period was insufficient to offset a 7-day exposure. Lanthanum exposure triggered a biochemical response as exposed clams showed increased SOD, CAT, GST, and TAC levels and lipoperoxidation occurred. The expression of heat shock proteins was majorly suppressed in La exposed clams, in combination of warming and acidification.

Hence, this study shows how the toxic effects of La may be enhanced in a changing world.

3.2.2. Echinoderms

Pagano et al. (2016) studied the effects of Y, La, Ce, Nd, Sm, Eu and Gd in sea urchin embryos and sperm (*Paracentrotus lividus*). Exposure from 10^{-6} to 10^{-4} M allured to increasing bioconcentration levels as: Gd > Y > La > Nd ≈ Eu > Ce ≈ Sm. Adverse effects were observed, regardless of exposure element. REE exposed embryos showed augmented MDA levels, ROS formation and NO levels. REE exposed sperm (10^{-5} to 10^{-4} M) caused concentration dependent fertilization impairment and offspring damage as decreased mitotic rate and higher anomaly rates. Gd, Y Ce and La lured the most toxic effects. Trifuoggi et al. (2017) assessed REE toxicity to early life stages of two distinct sea urchin species (*Sphaerechinus granularis* and *Arbacia lixula*). Concentrations from 10^{-7} to 10^{-4} M of Y, La, Ce, Nd, Sm, Eu and Gd were compared and the studied outpoints were developmental defects, cytogenetic anomalies, ferritization rate and offspring anomalies. Data showed distinct toxicity patterns between the studied elements. Moreover, following embryo and sperm exposure, *S. granularis* showed significantly higher sensitivity to *A. lixula* and *P. lividus* (Pagano et al., 2016). Here, the authors illustrate that individual REE elicit different toxicity patterns, in a gradient and, for example, La showed the highest toxicity to the developing embryos. Hence, a generalization of REE-induces toxicity may not be suitable.

3.2.3. Crustaceans

Barry and Meehan (2000) determined the acute and chronic toxicity of La to the crustacean *Daphnia carinata*. *Daphnia carinata* was exposed to 100, 200, 400, 600, 800, and 1000 µg L^{-1} La in three test media Daphnia growth media, hard water and carbon filtered and dechlorinated tap water. The authors showed that the La acute toxicity was sharply dependent on the media tested. In fact, La was most toxic to *D. carinata* in soft tap water, with 48-h EC50 of 43 µg L^{-1}, in comparison to 1180 µg L^{-1} in hard water and finally in daphnid growth medium (49 µg La L^{-1}), however, there was significant precipitation of La in this media. Although La caused delayed maturation, the assessment of mortality proved to be a more successful indicator of La toxicity, than growth or reproduction. In a much recent study, Korkmaz et al. (2021) explored the effect of Gd and La on the mortality of *Daphnia magna* by exposing the organisms to 10, 30, 50 and 100 mg L^{-1} La or Gd for 24, 48 and 72h. These authors described the highest mortality rate as 100% when *D.*

magna was exposed for 24h to 100 mg L^{-1} of both La and Gd. The same percentage of mortality was observed for the exposure at 50 and 100 mg L^{-1} of both La and Gd at 72h. Here La was also indicated as less toxic than Gd. Blinova et al. (2018) performed an acute toxicity study of La, Ce, Pr, Nd and Gd with freshwater crustaceans *Daphnia magna*, *Thamnocephalus platyurus* and *Heterocypris incongruens* and an additional chronic and recovery toxicity test with *D. magna*. The authors reasoned that due to methodical nuances the acute toxicity data of lanthanides is not reliable. After the 21-day exposure phase *D. magna* the mortality of parent dahpnids was more sensitive than reproduction. The LC50 values for the studied elements ranged from 0.3 to 0.5 mg L^{-1} and were not significantly different amongst them.

Phoslock ® is a lanthanum-modified clay applied to treat water bodies to trap excess dissolved phosphorus in the water column and released from sediments. Lürling and Tolman (2010) studied the impact of 33, 100, 330 and 1000 µg L^{-1} of Phoslock ®, in the absence and presence of phosphate (330 µg L^{-1}) and showed that in an environment without phosphorus La did not elicit toxic effects on the zooplankter grazer *Daphnia magna*. Nevertheless, when phosphorus was present, the formation of $LaPO_4$ resulted in considerable precipitation of algae which indirectly impacted life-history traits of this model. As the concentrations of La rose, when phosphorus was present, *D. magna* remained smaller, matured later, and reproduced less, resulting in lower population growth rates.

Bergsten-Torralba et al. (2020) assessed the toxicities of La^{3+}, Nd^{3+}, Sm^{3+}, and their combination to a variety of species such as the microalgae *Chlorella vulgaris* and *Raphidocelis subcapitata* the crustaceans *Daphnia similis* and *Artemia salina* and fungi *Penicillium simplicissimum* and *Aspergillus japonicus*. In this study La^{3+} was appointed as moderately toxic to microalgae and crustaceans. For *R. subcapitata* the 72h IC50 was 51.72 mg L^{-1} and for *C. vulgaris* the 72 h IC50 was 47.13 mg L^{-1}. Both microalgae showed a similar toxic response to lanthanum oxide and a significant lanthanum oxide precipitation occurred in all experiments, as it has been described for La chloride (Romero-Freire et al., 2019), nitrates, and oxides (Blinova et al., 2018). Here, *D. similis* was the most sensitive organism to La^{3+}. The mixtures, except $La^{3+}+Nd^{3+}$ were more toxic than the sum of their toxicities as single stressors. The freshwater crustacean *D. similis* was the most sensitive to La^{3+}, followed by both microalgae, and ultimately the seawater crustacean *A. salina*. The authors argued that medium salts may compete with La^{3+} for binding sites, or that La^{3+} levels are lowered by the formation of insoluble precipitates (i.e., carbonates and phosphates). This could be the case in a higher salinity environment, as carbonate species increases with an increase in ionic strength. Furthermore, it is known that salts may interfere with ions by forming complexes with anion, that present less bioavailability for organisms.

3.3. Plants

Several studies on terrestrial plants have appointed the REE as having low toxicity and even presenting beneficial effects at small concentrations, however this effect on aquatic plants is not straightforward. Aharchaou et al. (2020) studied the La and Ce toxicity to the unicellular green alga *Chlorella fusca*, as this species is known to grow on its reserves of phosphorus, even in phosphorus depleted medium. Uptake and toxicity of La and Ce were determined in and without 1×10^{-4} M of malic acid. For La, 2 synthetic ligands were tested: nitrilotriacetic acid (8×10^{-5} or 1×10^{-4} M) and iminodiacetic acid (1×10^{-4} M). Furthermore, to maintain a constant ionic strength, when the Ca concentration of the medium was increased by the addition of $Ca(NO_3)_2$, the concentration of KNO_3 was reduced and vice versa. Algal growth was measured every 24h throughout the 120h exposure phase by means of cell density, size distribution, and average surface area quantification. Here, La and Ce were described has presenting similar toxicity thresholds. Intracellular La and Ce concentration was correlated with inhibition of algal growth, despite the ligand. Increasing Ca concentrations, diminished the REE uptake, and vice-versa, which suggests that at a constant pH the REE uptake and toxicity depend on free ion concentration and ambient calcium availability. In fact, REE may compete among themselves or with other trivalent metals for binding sites. Nevertheless, the assessment of toxicity is tremendously complex due low solubility of REE, and the presence of hydroxides, phosphates, and carbonates.

Balusamy et al. (2015) studied the effect of La oxide3 nanoparticles on the freshwater microalgae *Chlorella* sp. and the crustacean *Daphnia magna*. For the algal growth inhibition assay, *Chlorella* sp. was exposed to La_2O_3 NP at 10, 50, 100, 250, 500 and 1000 mg L^{-1} and after 72h of exposure, the authors did not observe toxic effects. Nevertheless, EDS mapping showed La attached on the surface of this macroalgae, without provoking morphological modifications. At lower doses La enhanced growth, however, at higher doses, growth inhibition occurred. On the crustacean *D. magna*, significant effects were observed at the highest studied concentrations (effective concentration [EC50] 500 mg L^{-1}; lethal dose [LD50] 1000 mg L^{-1}). After 48h of exposure to La, 70% of mortality occurred with *D. magna* was exposed to 1000 mg L^{-1}. Hao et al. (1997) chose the algae *Chlorella vulgaris* to understand how REE species impact the bioconcentration processes. For such, the authors exposed this organism between 0.5 to 4 mg mL^{-1} of La(NO$_3$)$_3$, Gd$_2$O$_3$ and Y(NO$_3$)$_3$ and the ligands EDTA (ethylenediamine tetra acetic acid), NTA (nitrilotriacetic acid) and CIT (citrate), separately. The addition of organic ligands decreased sharply the bioaccumulation of La, Gd and Y. The bioconcentration order was La^{3+} > La-Cit > La-NTA > La-EDTA. As a 24h period until equilibrium was reached after exposure was observed, the authors suggested that the bioconcentration process included an initial phase of adsorption to cellular surfaced, that is followed of REE intracellular transport by diffusion or by a metabolic dependent process. Fujiwara et al. (2008) explored the toxicities of 33 metals, including La, to the microalgae *Chlorella kessleri* and described an IC50 of 313 µM La. Globally, metals with large ionic raddi present higher toxicity then lighter ions of the same chemical group, nevertheless, the authors could not find a universal parameter suitable to explain the studied elements toxicity. The authors also described that La^{3+} promotes the aggregation and precipitation of this microalgae cells, before fatal damage, suggesting that this may occur due to its high valence and large ionic radii. Sun et al. (2019) explored the effects of La on two bait algae (*Chlorella vulgaris* and *Phaeodactylum tricornutum*) through the assessment of photosynthetic changes and the antioxidant performance. The authors showed that La(NO$_3$)$_3$ impacted both species. In the first 48h of exposure to 2, 4, 6, 8, and 10 mg L^{-1} La(NO$_3$)$_3$, PS II was hampered, however values were restored to control values after 48h. The OJIP curve values (O, OJ, JI, and IP) decrease with the increasing concentrations of La. Nevertheless, exposure provoked oxidative stress, affecting algae growth.

Tai et al. (2010) intended to relate increasing availability of lanthanides in seawater to biological toxicity on a marine monocellular algae (*Skeletonema costatum*). The REE exposure concentrations reached 5, 10, 20, 30, and 40 µmol L^{-1}, except Y. The results show that every REE provoked comparable toxic effects. High concentrations (29.04 ± 0.61 µmol L^{-1}) caused a growth reduction of 50%, after 96h (EC50). Here, the "Harkins rule" was not suitable. A solution with equivalent concentrations of each REE caused the same inhibition effect as each individual element at the same total concentration, which is a unique chemical phenomenon. The authors suggested that monocellular algae could not proficiently differentiate between these chemically coherent group.

Siciliano et al. (2021b) exposed the microalgae *Raphidocelis subcapitata* to La and Ce for 3 days, to understand the effective concentration inhibiting growth on 10%. EC10 was 0.5 mg L^{-1} for La and 0.4mg L^{-1} for Ce. These values were then used as nominal exposure concentrations, for 28 days and the organism was sampled after 7, 14, 21 and 28 days. The studied elements were bioaccumulated up to 5-fold. Both La and Ce enhanced growth inhibition up to 38 and 28%, respectively. Reactive oxygen species, superoxide dismutase and catalase revealed distinct effects from day 7 to 14, 21 and 28. Globally, La and Ce effects appeared to be counteracted. The authors of this study discussed that while they presented evidence on the ability of a primary producer to bioconcentrate La and Ce from water, information in the literature is still lacking on the biomagnification potential. Siciliano et al. (2021a) explored the effect of concentrations between 0.01 and 10.22 mg L^{-1} of Ce, Gd, La and Nd to simulate acid mine drainage, at pH of 4 and 6 on terrestrial plants, namely *Vicia faba*, *Lepidium sativum* and on the microalgae *Raphidocelis subcapitata*. Speciation analysis showed that the studied elements were mainly present as trivalent ions (86-88%) at pH 4. Gd, La and Nd were more toxic at pH 4, when the percentage of trivalent ions was greater. At pH 4, the toxicity trend was La < Ce < Nd < Gd and this trend was altered at pH 6: Nd < Ce < Gd ≈ La. Lanthanum, Gd and Nd caused mutagenicity at pH 4. Regarding the studied microalgae species, the REE ecotoxicity did not always increase with increasing atomic number. The authors suggest that possible negative effects may derive from La exposure in an environment impacted by acid mine drainage, however this may be decreased by altering pH.

Ashraf et al. (2021) explored the toxic effects of La on the freshwater microalga *Desmodesmus quadricauda* and analyzed the intracellular location of La with μXRF tomography. Increasing levels of La triggered cell count decline and inhibition of photosynthesis. Metalloproteomics revealed that at sublethal concentration (128 nM) La was in hotspots inside the cells, while at lethal 1387 nM that led to release of other ions (K, Zn) from the cells, La filled most of the cells. La reduced photochemistry at nanomolar levels, and the decreased captured energy elicited reduced cell division, as showcased by cell density measurements.

Toxic strains of algal blooms produce microcystins (MCs), that may compromise human and ecosystem health. As previously discussed before, La and the remaining lanthanides may be accumulated by algae and cause growth and physiological impairments. Nevertheless, the impacts of REE on cellular macrocystins was not known. Therefore, Shen et al. (2018) explored the effects of La in *Microcystis aeruginosa,* a freshwater cyanobacteria, through stoichiometry. *M. aeruginosa* was exposed to 0.2, 2.0, 4.0, 8.0, 20, 40, and 60 µM of La. Low dose exposure increased MCs, microcystin-YR, microcystin-LW content and decreased microcystin-LR. High doses (8.0 µM and above) reduced total content of MCs, microcystin-YR and microcystin-LW while increasing content of microcystin-LR. Alterations in MCs content were associated with the ratios C:P and N:P in algal cells. The composition of MCs was associated to the ratio of C:N. Therefore, La may impact the content and composition of MCs, through altering the growth and chlorophyll-a content of *Microcystis aeruginosa*, which may ultimately influence Microcystis blooms toxicity. Song et al. (2021) applied a Magnetite/lanthanum hydroxide composite (MLC) in a water, sediment, and cyanobacteria system, to understand the effects of this magnetic La-based material on cyanobacterial bloom. The exposed system to 0.5 g and 1.0 g of MLC diminished the cyanobacteria density in 99.39% and 99.84%, after 30 days. The authors reasoned that this result could be related to the high adsorption of MLC to phosphorus. In fact, MLC may form a phosphorus-binding layer that adsorbed soluble reactive phosphate (SRP), and enhanced internal phosphorus (P) release from sediment and obstructed soluble reactive phosphate release from pore water. Exposition to MLC elicited oxidative stress and damage, and this damage with greater with increasing exposure concentration. The authors proposed the application of MLC as a P-inactive material for cyanobacterial bloom controlling.

Jin et al. (2009) explored the effects of La and EDTA on the growth competition of the cyanobacterium *Microcystis aeruginosa* and the green alga *Scenedesmus quadricauda*. In this experiment, the concentration of Fe-citrate was decreased to 3 µmol L^{-1} to reduce the influence of citrate on the speciation of lanthanum. The effects of Lanthanum at 0, 0.72, 7.2 and 72 µmol L^{-1} (0, 0.1,1 and 10 mg L^{-1}) were investigated, and the concentrations of EDTA varied from 0.269 to 26.9 µmol L^{-1}. With low doses of EDTA (0.269 µmol L^{-1}), the low La levels (7.2 µmol L^{-1}) showed little stimulative growth effects on *M. aeruginosa* and *S. quadricauda*. On another hand, high La levels (72 µmol L^{-1}) showed a significant growth inhibition on both studied species. Cultivation experiments showed higher inhibitory impacts on *M. aeruginosa* than *S. quadricauda*, and hence, in mixed cultures the former species could dominate the first. Instead, without La addition, high levels of EDTA (413.4 µmol L^{-1}) inhibited *M. aeruginosa* growth and showed slight effects on *S. quadricauda*. The presence of EDTA lessened the La induced growth inhibition on *M. aeruginosa* at intermediate levels (2.69-13.4 µmol L^{-1}).

Regarding macroalgae, Costa et al. (2020) studied the influence of REE (lanthanum, cerium, praseodymium, neodymium, europium, gadolinium, terbium, dysprosium, and yttrium) and salinity (10 and 30) on the removal of potentially toxic elements (Cd, Cr, Cu, Hg, Ni and Pb) by different species (*Fucus spiralis*, *Fucus vesiculosus*, *Gracilaria* sp., *Osmundea pinnatifida*, *Ulva intestinalis* and *Ulva lactuca*). The macroalgae were exposed to 1μmol L^{-1} of every element for 7 days, at both salinities. No effects on uptake of Ni and Hg were observed. For the remaining elements, the effect of the presence of REE was species specific. The authors suggest that such differences may be related by REE chemistry in water and by the distinct algae physiological traits. As species were exposed at the same time to every element, competition could have occurred, however this study illustrated the capacity of these macroalgae on diminishing the levels of potentially toxic elements, even when the poorly understood emerging REE are present. Figueiredo et al. (2022d) investigated the *Ulva rigida* phytoremediation capacity and ecotoxicological responses after exposure to 15 μg L^{-1} La and 10 μg L^{-1} Gd, separately, in a warming and acidification setting. A 7-day exposure phase was followed by a 7-day elimination. Warming and Acidification contributed to the lowest La accumulation, and increased elimination. Lanthanum and Gd activated the antioxidant defence system, and lipid damage was avoided. Nonetheless, La and Gd exposure in climate change scenarios requested an enhanced antioxidant response and a total chlorophyll and carotenoids rise which may imply a superior energy expenditure, as a response to a multi-stressor near-future environment.

Weltje et al. (2002) described La accumulation and elimination by exponentially growing common duckweed (*Lemna minor* L.). Duckweed was exposed to 10 nM of La and the authors observed that 25% of the La present was adsorbed on glass walls of the experimental tanks. Growth-induced dilution seemed more effective than elimination itself. Average bioconcentration facter was 2000 L Kg^{-1} ww (30000 L Kg^{-1} dw), which is equivalent to other higher plants. At this exposure concentration, no effects on growth occurred. Ippolito et al. (2010) investigated the impacts of a mix of REE nitrates and La nitrate on the biochemical performance associated with the ascorbate–glutathione cycle in the common duckweed *Lemna minor*. A La nitrate was selected for this study as this is the form present in REE-enriched fertilizers. The duckweed was incubated for 2 and 5 days with 1, 5, and 10 mM of La and 1, 5, and 10 mM of REE (100.07 mM La, 327.57 mM Ce, 25.76 mM Pr, 0.14 mM Nd and 0.006 mM Gd). Concentrations bellow 5 mM of REE and La did not impact ROS production, chlorophyll content, and lipid peroxidation. Toxicity occurred after 5 days of exposition to 10 mM of a mix of REE and La. Increased glutathione and enzymatic antioxidants occurred before stress symptoms in exposed duckweed was observed. Exposed duckweed plants were also exposed to chilling stress. Although *L. minor* is known to be tolerant to cold, preexposure to REE and La diminished its thermal tolerance.

Xu et al. (2012) explored the structural effects of La (up to 160 µM) in the aquatic plant *Hydrocharis dubia*. Lanthanum accumulation was concentration dependent, and this element disrupted the balance of key nutrient elements (P, Mg, Ca, Fe, K, and Zn). The EC50 for chlorophyll was 20 µM on day 7 and pigment content diminished with exposure to La. Although superoxide dismutase, peroxidase, catalase, reduced ascorbate, and reduced glutathione showed distinct response patters at distinct La exposure concentrations, malondialdehyde increased consistently with increasing La concentrations. Overall, damage to chloroplast, mitochondrion, and nucleus was observed, which could indicate that La may disrupt normal cellular functions in this plant species.

Conclusion

Many studies described here have been performed with only one or with a small number of elements, though the Lanthanides are naturally found as a whole group. The booming number of published articles on La toxicity in aquatic environments is yet far from being a match to the usage rise of REE, their developing industrial applications and consequent environmental discharge, with impacts to biodiversity, ecosystem functioning and human economy and health. Together with open research questions regarding La, and the remaining Lanthanides, uptake and elimination routs and toxic effects, this gap should be further lessened in future studies. Furthermore, it is of the uttermost significance to better understand the natural REE levels in pristine and impacted water bodies to improve the knowledge on REE bioavailability and better design environmentally realistic exposure experiments. Although the REE availability is changing, the assessment of its behavior, uptake, and toxic effects together with other abiotic factors that are also changing is almost inexistent. Although several other contaminants have been widely studied in near future scenarios, only a couple of studies have detailed the behavior of these elements on a climate change perspective (i.e., ocean warming, acidification, and salinity changes).

The results gathered in this chapter show that La may prompt stimulatory and toxic effects, depending on the exposure concentration and the studied model species.

The assessment of chemical speciation has been shown to be critical, however, only a small number of studies has accounted for the possibility of the exposure concentrations being much lower than nominal desired ones due to the formation of insoluble chemical species. Unfortunately, data on environmental parameters and speciation are frequently not detailed in publications.

Although the number of studies has increased in recent years, the Lanthanum modes of action remain poorly understood. Through these modes, the trivalent La^{3+} ia able to cross cellular membranes and might compete with Ca^{2+} for binding sites, impairing calcium channels and cell membranes transport with consequences to cell functioning. As La-uptake is impacted by a wide array of factors, considerable inconsistent La bioaccumulation patterns have been described. Metal speciation, environmental physicochemical parameter and physiological characteristics of the studied organisms have been shown to interfere with La bioaccumulation. Furthermore, La was bioaccumulated as a dissolved free ion and complexed.

As of now, records show that La may elicit indirect and direct toxic effects. For example, various authors described that the impacts on algae growth could be related to phosphate binding and therefore limitation, rather than direct effects from La uptake. Numerous knowledge gaps about La ecotoxicity and uptake by various aquatic organism remain, this include the ability of this element to be transferred along the food web. La exposure may engender reactive oxygen species, lipid peroxidation, and modulate anti-oxidation activities. Organisms' adaptation to increasing concentrations of La and their ability to detoxify it are to this date also poorly described. REEs' safety levels remain to be determined.

The lack of regulations for REE discharge into the environment through the domestic, agricultural, or industrial sector, accoupled with inefficient and ecologically inadequate treatment methods and disposal and recycle technologies of e-waste and other modern life every-day items containing significant REEs levels, highlight the urgent need for the scientific community to focus its efforts on this recent emergent and yet poorly understood problematic.

Conflict of Interest

The authors declare no conflict of interests.

Acknowledgments

This work was supported by the European Union's operation program Mar 2020 through the research project CEIC (MAR-01.04.02-FEAMP-0012) awarded to Joana Raimundo.

References

Aharchaou, I., Beaubien, C., Campbell, P. G., Fortin, C., 2020. Lanthanum and cerium toxicity to the freshwater green alga Chlorella fusca: applicability of the biotic ligand model. *Environmental Toxicology and Chemistry* 39, 996-1005. doi: 10.1002/etc.4707.

Andrade, M., Soares, A. M., Solé, M., Pereira, E., Freitas, R., 2021. Salinity influences on the response of Mytilus galloprovincialis to the rare-earth element lanthanum. *Science of the Total Environment* 794, 148512. doi: 10.1016/j.scitotenv.2021.148512.

Ashraf, N., Vítová, M., Cloetens, P., Mijovilovich, A., Bokhari, S. N. H., Küpper, H., 2021. Effect of nanomolar concentrations of lanthanum on Desmodesmus quadricauda cultivated under environmentally relevant conditions. *Aquatic Toxicology* 235, 105818.

Atwood, D. A., 2013. *The rare earth elements: fundamentals and applications*. John Wiley & Sons. doi: 10.1016/j.aquatox.2021.105818.

Balaram, V., 2019. Rare earth elements: A review of applications, occurrence, exploration, analysis, recycling, and environmental impact. *Geoscience Frontiers* 10, 1285-1303. doi: 10.1016/j.gsf.2018.12.005.

Balusamy, B., Taştan, B. E., Ergen, S. F., Uyar, T., Tekinay, T., 2015. Toxicity of lanthanum oxide (La 2 O 3) nanoparticles in aquatic environments. *Environmental Science: Processes & Impacts* 17, 1265-1270. doi: 10.1039/C5EM00035A.

Barry, M. J., Meehan, B. J., 2000. The acute and chronic toxicity of lanthanum to Daphnia carinata. *Chemosphere* 41, 1669-1674. doi: 10.1016/S0045-6535(00)00091-6.

Bergsten-Torralba, L. R., Magalhães, D. d. P., Giese, E. C., Nascimento, C., Pinho, J., Buss, D., 2020. Toxicity of three rare earth elements, and their combinations to algae, microcrustaceans, and fungi. *Ecotoxicology and Environmental Safety* 201, 110795. doi: 10.1016/j.ecoenv.2020.110795.

Binnemans, K., McGuiness, P., Jones, P. T., 2021. Rare-earth recycling needs market intervention. *Nature Reviews Materials* 6, 459-461. doi: 10.1038/s41578-021-00308-w.

Birge, W., Black, J., Westerman, A., 1979. Evaluation of aquatic pollutants using fish and amphibian eggs as bioassay organisms. *Animals as monitors of environmental pollutants/sponsored by Northeastern Research Center for Wildlife Diseases, University of Connecticut, Registry of Comparative of Lab Animal Resources, National Academy of Sciences.*

Blinova, I., Lukjanova, A., Muna, M., Vija, H., Kahru, A., 2018. Evaluation of the potential hazard of lanthanides to freshwater microcrustaceans. *Science of the Total Environment* 642, 1100-1107. doi: 10.1016/j.scitotenv.2018.06.155.

Burch, K. R., Comer, J. B., Wolf, S. F., Brake, S. S., 2011. REE geochemistry of an acid mine drainage system in western Indiana, Abstr. with Programs—*Geol. Soc. Am*, p. 587p.

Carpenter, D., Boutin, C., Allison, J. E., Parsons, J. L., Ellis, D. M., 2015. Uptake and effects of six rare earth elements (REEs) on selected native and crop species growing in contaminated soils. *PloS One* 10, e0129936. doi: 10.1371/journal.pone.0129936.

Chen, Y., Cao, X., Lu, Y., Wang, X., 2000. Effects of rare earth metal ions and their EDTA complexes on antioxidant enzymes of fish liver. *Bulletin of environmental contamination and toxicology* 65, 357-365. doi: 10.1007/s001280000136.

Costa, M., Henriques, B., Pinto, J., Fabre, E., Viana, T., Ferreira, N., Amaral, J., Vale, C., Pinheiro-Torres, J., Pereira, E., 2020. Influence of salinity and rare earth elements on simultaneous removal of Cd, Cr, Cu, Hg, Ni and Pb from contaminated waters by living macroalgae. *Environmental Pollution* 266, 115374. doi: 10.1016/j.envpol.2020.115374.

Cunha, M., Louro, P., Silva, M., Soares, A. M., Pereira, E., Freitas, R., 2022. Biochemical alterations caused by lanthanum and gadolinium in Mytilus galloprovincialis after exposure and recovery periods. *Environmental Pollution* 307, 119387. doi: 10.1016/j.envpol.2022.119387.

Dubé, M., Auclair, J., Hanana, H., Turcotte, P., Gagnon, C., Gagné, F., 2019. Gene expression changes and toxicity of selected rare earth elements in rainbow trout juveniles. *Comparative Biochemistry and Physiology Part C: Toxicology & Pharmacology* 223, 88-95. doi: 10.1016/j.cbpc.2019.05.009.

Eddy, F., Bath, R., 1979. Effects of lanthanum on sodium and chloride fluxes in the goldfish Carassius auratus. *Journal of comparative physiology* 129, 145-149.

Figueiredo, C., Grilo, T. F., Lopes, A. R., Lopes, C., Brito, P., Caetano, M., Raimundo, J., 2022a. Differential tissue accumulation in the invasive Manila clam, Ruditapes philippinarum, under two environmentally relevant lanthanum concentrations. *Environmental monitoring and assessment* 194, 1-11.

Figueiredo, C., Grilo, T. F., Lopes, C., Brito, P., Caetano, M., Raimundo, J., 2022b. Lanthanum and Gadolinium availability in aquatic mediums: New insights to ecotoxicology and environmental studies. *Journal of Trace Elements in Medicine and Biology* 71, 126957.

Figueiredo, C., Grilo, T. F., Lopes, C., Brito, P., Diniz, M., Caetano, M., Rosa, R., Raimundo, J., 2018. Accumulation, elimination and neuro-oxidative damage under lanthanum exposure in glass eels (Anguilla anguilla). *Chemosphere* 206, 414-423.

Figueiredo, C., Grilo, T. F., Oliveira, R., Ferreira, I. J., Gil, F., Lopes, C., Brito, P., Ré, P., Caetano, M., Diniz, M., 2022c. Single and combined ecotoxicological effects of ocean warming, acidification and lanthanum exposure on the surf clam (Spisula solida). *Chemosphere* 302, 134850.

Figueiredo, C., Grilo, T. F., Oliveira, R., Ferreira, I. J., Gil, F., Lopes, C., Brito, P., Ré, P., Caetano, M., Diniz, M., 2022d. A triple threat: ocean warming, acidification and rare earth elements exposure triggers a superior antioxidant response and pigment production in the adaptable Ulva rigida. *Environmental Advances*, 100235.

Figueiredo, C., Raimundo, J., Lopes, A. R., Lopes, C., Rosa, N., Brito, P., Diniz, M., Caetano, M., Grilo, T. F., 2020. Warming enhances lanthanum accumulation and toxicity promoting cellular damage in glass eels (Anguilla anguilla). *Environmental research* 191, 110051.

Fujiwara, K., Matsumoto, Y., Kawakami, H., Aoki, M., Tuzuki, M., 2008. Evaluation of metal toxicity in Chlorella kessleri from the perspective of the periodic table. *Bulletin of the Chemical Society of Japan* 81, 478-488.

Gonzalez, V., Vignati, D. A., Leyval, C., Giamberini, L., 2014. Environmental fate and ecotoxicity of lanthanides: are they a uniform group beyond chemistry? *Environment International* 71, 148-157.

Gwenzi, W., Mangori, L., Danha, C., Chaukura, N., Dunjana, N., Sanganyado, E., 2018. Sources, behaviour, and environmental and human health risks of high-technology rare earth elements as emerging contaminants. *Science of the Total Environment* 636, 299-313.

Hanana, H., Kleinert, C., Gagné, F., 2021. Toxicity of representative mixture of five rare earth elements in juvenile rainbow trout (Oncorhynchus mykiss) juveniles. *Environmental Science and Pollution Research* 28, 28263-28274.

Hao, S., Xiaorong, W., Liansheng, W., Lemei, D., Zhong, L., Yijun, C., 1997. Bioconcentration of rare earth elements lanthanum, gadolinium and yttrium in algae (Chlorella Vulgarize Beijerinck): Influence of chemical species. *Chemosphere* 34, 1753-1760.

Hazra, A., Das, S., Ganguly, A., Das, P., Chatterjee, P., Murmu, N., Banerjee, P., 2019. Plasma arc technology: a potential solution toward waste to energy conversion and of ghgs mitigation, *Waste valorisation and recycling*. Springer, pp. 203-217.

Herrmann, H., Nolde, J., Berger, S., Heise, S., 2016. Aquatic ecotoxicity of lanthanum–A review and an attempt to derive water and sediment quality criteria. *Ecotoxicology and Environmental Safety* 124, 213-238.

Hu, B., He, M., Jakubowski, N., Meinhardt, J., Meyer, F. M., Niederstraßer, J., Schramm, R., Sindern, S., Stosch, H.-G., Bertau, M., 2017. *Handbook of Rare Earth Elements: Analytics*. Walter de Gruyter GmbH & Co KG.

Hua, D., Wang, J., Yu, D., Liu, J., 2017. Lanthanum exerts acute toxicity and histopathological changes in gill and liver tissue of rare minnow (Gobiocypris rarus). *Ecotoxicology* 26, 1207-1215.

Huang, Z., Gao, N., Zhang, S., Xing, J., Hou, J., 2022. Investigating the toxically homogenous effects of three lanthanides on zebrafish. *Comparative Biochemistry and Physiology Part C: Toxicology & Pharmacology* 253, 109251.

Humphries, M., 2010. *Rare earth elements: the global supply chain*. Diane Publishing.

Ippolito, M., Fasciano, C., d'Aquino, L., Morgana, M., Tommasi, F., 2010. Responses of antioxidant systems after exposition to rare earths and their role in chilling stress in common duckweed (Lemna minor L.): a defensive weapon or a boomerang? *Archives of environmental contamination and toxicology* 58, 42-52.

Jin, X., Chu, Z., Yan, F., Zeng, Q., 2009. Effects of lanthanum (III) and EDTA on the growth and competition of Microcystis aeruginosa and Scenedesmus quadricauda. *Limnologica* 39, 86-93.

Korkmaz, V., Ergüven, G. Ö., Yıldırım, N., Yıldırım, N. C., 2021. The Effect of Gadolinium and Lanthanum on the Mortality of Daphnia magna. Uluslararası tarım araştırmalarında yenilikçi yaklaşımlar dergisi [*International journal of innovative approaches in agricultural research*] (Online) 5, 213-220.

Kulaksız, S., Bau, M., 2011. Rare earth elements in the Rhine River, Germany: first case of anthropogenic lanthanum as a dissolved microcontaminant in the hydrosphere. *Environment International* 37, 973-979.

Liang, P., Liu, Y., Guo, L., 2005. Determination of trace rare earth elements by inductively coupled plasma atomic emission spectrometry after preconcentration with multiwalled carbon nanotubes. *Spectrochimica Acta Part B: Atomic Spectroscopy* 60, 125-129.

Liu, D., Yu, H., Gu, Y., Pang, Q., 2021. Effect of rare earth element lanthanum on lipid deposition and Wnt10b signaling in the liver of male zebrafish. *Aquatic Toxicology* 240, 105994.

Luoma, S. N., Rainbow, P. S., 2005. Why is metal bioaccumulation so variable? Biodynamics as a unifying concept. *Environmental science & technology* 39, 1921-1931.

Lürling, M., Tolman, Y., 2010. Effects of lanthanum and lanthanum-modified clay on growth, survival and reproduction of Daphnia magna. *Water research* 44, 309-319.

Malhotra, N., Hsu, H.-S., Liang, S.-T., Roldan, M. J. M., Lee, J.-S., Ger, T.-R., Hsiao, C.-D., 2020. An updated review of toxicity effect of the rare earth elements (REEs) on aquatic organisms. *Animals* 10, 1663.

Martin, M., Hickey, C., 2004. *Determination of HSNO ecotoxic thresholds for granular Phoslock™ (Eureka 1 formulation) phase 1: acute toxicity*. Report prepared for Primaxa Ltd, 2004-2137.

Mestre, N. C., Sousa, V. S., Rocha, T. L., Bebianno, M. J., 2019. Ecotoxicity of rare earths in the marine mussel Mytilus galloprovincialis and a preliminary approach to assess environmental risk. *Ecotoxicology* 28, 294-301.

Moreira, A., Henriques, B., Leite, C., Libralato, G., Pereira, E., Freitas, R., 2020. Potential impacts of lanthanum and yttrium through embryotoxicity assays with Crassostrea gigas. *Ecological Indicators* 108, 105687.

Pagano, G., Guida, M., Siciliano, A., Oral, R., Koçbaş, F., Palumbo, A., Castellano, I., Migliaccio, O., Thomas, P. J., Trifuoggi, M., 2016. Comparative toxicities of selected rare earth elements: Sea urchin embryogenesis and fertilization damage with redox and cytogenetic effects. *Environmental research* 147, 453-460.

Pinto, J., Costa, M., Leite, C., Borges, C., Coppola, F., Henriques, B., Monteiro, R., Russo, T., Di Cosmo, A., Soares, A. M., 2019. Ecotoxicological effects of lanthanum in Mytilus galloprovincialis: Biochemical and histopathological impacts. *Aquatic Toxicology* 211, 181-192.

Qiu, Y., Hua, D., Liu, J., Wang, J., Hu, B., Yu, D., 2020. Induction of micronuclei, nuclear anomalies, and dimensional changes in erythrocytes of the rare minnow (Gobiocypris rarus) by lanthanum. *Environmental Science and Pollution Research* 27, 31243-31249.

Rim, K.-T., 2016. Effects of rare earth elements on the environment and human health: a literature review. *Toxicology and Environmental Health Sciences* 8, 189-200.

Romero-Freire, A., Joonas, E., Muna, M., Cossu-Leguille, C., Vignati, D., Giamberini, L., 2019. Assessment of the toxic effects of mixtures of three lanthanides (Ce, Gd, Lu) to aquatic biota. *Science of the Total Environment* 661, 276-284.

Shen, F., Wang, L., Zhou, Q., Huang, X., 2018. Effects of lanthanum on Microcystis aeruginosa: Attention to the changes in composition and content of cellular microcystins. *Aquatic Toxicology* 196, 9-16.

Siciliano, A., Guida, M., Pagano, G., Trifuoggi, M., Tommasi, F., Lofrano, G., Suarez, E. G. P., Gjata, I., Brouziotis, A. A., Liguori, R., 2021a. Cerium, gadolinium, lanthanum, and neodymium effects in simplified acid mine discharges to Raphidocelis subcapitata, Lepidium sativum, and Vicia faba. *Science of the Total Environment* 787, 147527.

Siciliano, A., Guida, M., Serafini, S., Micillo, M., Galdiero, E., Carfagna, S., Salbitani, G., Tommasi, F., Lofrano, G., Suarez, E. G. P., 2021b. Long-term multi-endpoint exposure of the microalga Raphidocelis subcapitata to lanthanum and cerium. *Science of the Total Environment* 790, 148229.

Song, Q., Huang, S., Xu, L., Li, Q., Luo, X., Zheng, Z., 2021. Response of Magnetite/Lanthanum hydroxide composite on cyanobacterial bloom. *Chemosphere* 275, 130017.

Sun, D., He, N., Chen, Q., Duan, S., 2019. Effects of lanthanum on the photosystem II energy fluxes and antioxidant system of Chlorella Vulgaris and Phaeodactylum Tricornutum. *International Journal of Environmental Research and Public Health* 16, 2242.

Tai, P., Zhao, Q., Su, D., Li, P., Stagnitti, F., 2010. Biological toxicity of lanthanide elements on algae. *Chemosphere* 80, 1031-1035.

Trifuoggi, M., Pagano, G., Guida, M., Palumbo, A., Siciliano, A., Gravina, M., Lyons, D. M., Burić, P., Levak, M., Thomas, P. J., 2017. Comparative toxicity of seven rare earth elements in sea urchin early life stages. *Environmental Science and Pollution Research* 24, 20803-20810.

Uchida, N., Matsukami, H., Someya, M., Tue, N. M., Viet, P. H., Takahashi, S., Tanabe, S., Suzuki, G., 2018. Hazardous metals emissions from e-waste-processing sites in a village in northern Vietnam. *Emerging Contaminants* 4, 11-21.

Weltje, L., Brouwer, A. H., Verburg, T. G., Wolterbeek, H. T., de Goeij, J. J., 2002. Accumulation and elimination of lanthanum by duckweed (Lemna minor L.) as influenced by organism growth and lanthanum sorption to glass. *Environmental Toxicology and Chemistry: An International Journal* 21, 1483-1489.

Xu, Q., Fu, Y., Min, H., Cai, S., Sha, S., Cheng, G., 2012. Laboratory assessment of uptake and toxicity of lanthanum (La) in the leaves of Hydrocharis dubia (Bl.) Backer. *Environmental Science and Pollution Research* 19, 3950-3958.

Chapter 3

Synthesis and Properties of Nanodispersed Luminescent Structures Based on Lanthanum Fluoride and Phosphate for Optopharmacology and Photodynamic Therapy of Tumor Diseases Localized in Cranial Organs and Bone Tissues

A. Kusyak[1], A. Petranovska[1], O. Oranska[1],
S. Turanska[1], Ya. Shuba[2], D. Kravchuk[2],
L. Kravchuk[2], G. Sotkis[2], V. Nazarenko[3],
R. Kravchuk[3], V. Dubok[4], O. Bur'yanov[5],
V. Chornyi[5], Yu. Sobolevs'kyy[5] and P. Gorbyk[1,*]

[1]Department of Nanomaterials, Chuiko Institute of Surface Chemistry of NAS of Ukraine, Kyiv, Ukraine
[2]Department of Neuromuscular Physiology, Bogomolets Institute of Physiology of NAS of Ukraine, Kyiv, Ukraine
[3]Department of Physics of Crystals, Institute of Physics of NAS of Ukraine, Kyiv, Ukraine
[4]Department of Solids Structural Chemistry, Frantsevich Institute of Problems of Materials Science, NAS of Ukraine, Kyiv, Ukraine
[5]Department of Traumatology and Orthopedics of the National Medical University (named after O. O. Bohomolets), Kyiv, Ukraine

---

[*] Corresponding Author's Email: phorbyk@ukr.net.

In: What to Know about Lanthanum
Editor: Catherine C. Bradley
ISBN: 979-8-88697-615-1
© 2023 Nova Science Publishers, Inc.

Abstract

The aim of the work is the synthesis of nanodispersed phosphors based on lanthanum fluoride and lanthanum phosphate activated by terbium ($LaF_3:Tb^{3+}$ and $LaPO_4:Tb^{3+}$, respectively), promising for use in photodynamic therapy and optopharmacology, study of their structural properties and luminescence as well as the possibility of their use in nanocomposites (NC) with magnetically sensitive nanosized Fe_3O_4 carriers and 60S bioactive glass. Terbium-activated nanocrystalline lanthanum fluoride and lanthanum phosphate of hexagonal syngony were synthesized. Structural properties, chemical activity of surface, UV and X-ray luminescence spectra of the synthesized crystals have been studied. The possibility is shown to use them in NC with magnetically sensitive nanosized drug carriers and bioactive sol-gel glass. The acid-base nature of the active surface centers of LaF_3 and $LaF_3:Tb^{3+}$ NPs was determined. The parts of negatively α^-, positively α^+ charged and neutral α^0 active centers were calculated in the pH range of 2.4 – 12.7. The detected active centers of the surface can be represented by acidic (La^{3+}) and base (F^-) Lewis centers, as well as base Bronsted centers (OH^- groups). The obtained data are useful for optimization of the conditions of adsorption immobilization of molecules of photosensitive substances (photosensitizers) from physiological solution onto the surface of phosphors based on lanthanum fluoride. Ensembles of particles of magnetically sensitive NC $Fe_3O_4/LaF_3:Tb^{3+}$ of the core-shell type were synthesized. Conditions for the synthesis of NC did not significantly change the magnetic properties of their cores – the original single-domain Fe_3O_4 nanoparticles. 60S BG composites with nanodispersed crystalline $LaF_3:Tb^{3+}$ and $LaPO_4:Tb^{3+}$ in the dry state, and in distilled water, showed the presence of luminescence upon excitation by UV and X-rays. These data indicate the potential of research into nanodispersed phosphors based on lanthanum fluoride and lanthanum phosphate, their composites with magnetically sensitive nanosized carriers and bioactive glass, for use in optopharmacology and photodynamic therapy of tumor diseases localized in cranial organs and bone tissues. In addition, the results of research can be useful for technical applications, in particular, in the creation of luminescent detectors of high-energy electromagnetic radiation, the development of photo- and optoelectronic devices, etc.

Keywords: lanthanum fluoride, lanthanum phosphate, luminescent nanostructures, magnetite, nanocomposites, optopharmacology, photodynamic therapy, sol-gel glass

Introduction

The creation of effective multifunctional drugs for targeted delivery and local therapy with prolonged release of bioactive components is an urgent problem in many fields of modern medicine (Roco, Williams, and Alivisatos, 2002; Gorobets', Gorobets', Gorbyk et al., 2018). Thus, for oncology, the concept of chemical design of magnetosensitive nanocomposites (NCs) of core-shell type (Abramov, Turanska, and Gorbyk, 2018a; Abramov, Turanska, and Gorbyk, 2018b) with multilevel hierarchical layered shell nanoarchitecture has become a priority. These NCs are able to perform the functions of medico-biological nanorobots (Shpak and Gorbyk, 2009; Gorbyk, Lerman, Petranovska et al., 2016; Abramov, Kusyak, Kaminskiy et al., 2017): to recognize specific cells, viruses and biomacromolecules in biological media; perform the functions of targeted delivery and accumulation of drugs in target cells and organs; complex local chemo-, immuno-, neutron-capture-, hyperthermic-, photodynamic therapy (PDT) and magnetic resonance imaging diagnostics in real time; detoxification of the body by adsorption of degradation residues of cells, viral particles, heavy metal ions, etc. and their removal by magnetic field. The scheme of chemical design of NCs with the functions of nanorobots is given in (Gorbyk, Lerman, Petranovska et al., 2016).

The optimal way of practical application of magnetically sensitive multifunctional NCs in medicine (Gorbyk, 2020) may be the creation of magnetic fluids based on them and saline (Abramov, Turanska, and Gorbyk, 2018a; Abramov, Turanska, and Gorbyk, 2018b). In particular, in the composition of magnetic fluids, such NCs can be used for treatment of tumor diseases localized in cranial organs, for example, by PDT modified to work with photosensitizers, exciting them by high-permeability "soft" X-ray radiation, safe for patients (Min-Hua, Yi-Jhen, Sheng-Kai et al., 2017).

In onco-orthopedic surgery, the priority is to develop new types of implants for use as a complex delivery system of chemotherapeutic and osteoconductive drugs with prolonged action for local use (Kusyak, Petranovska, Dubok et al., 2021; Kusyak, Petranovska, Turanska et al., 2021). In this direction, bioactive ceramics, in particular, different types of sol-gel glass have an undeniable advantage over many other drug carriers, because they are biocompatible, do not cause a negative immune response, quickly and securely fixed due to direct biochemical interaction with adjacent tissues, not encapsulated with the formation of connective tissue (which is characteristic of foreign materials), as well as they biodegrade gradually in the body by resorption and biochemical reactions (Hench and Fielder, 2004; Dutra, Pereira, Serakides et al., 2008). A significant advantage of bioceramic materials is the ability to introduce into their composition the necessary substances in order to expand their functional properties and increase efficiency (Buryanov, Chornyi, Protsenko et al., 2018; Buryanov, Chornyi, Dedukh et al., 2019). In particular, in recent years, the efforts of researchers in this field are aimed at implementation of PDT method, which can be used for minimally invasive treatment of malignant tumorous formations localized both in soft tissues and in bone structures, when using biocompatible materials with X-ray luminescent properties, as well as effective photosensitizers (Liu, Chen, Wang et al., 2008; Hsiu-Wen, Chien-Hao, Chien-Hsin et al., 2020; Medkov, Steblevskaya, and Belobeletskaya, 2017).

Thus, a promising approach that can provide minimally invasive treatment of malignant tumors localized in organs of crane, bone tissues, etc., is to create a new optopharmacological base for PDT based on the use of magnetically sensitive carriers for targeted drug delivery (Mangaiyarkarasi, Chinnathambi, Karthikeyan et al., 2016; Zhang, Braun, Pallaoro et al., 2011; DiMaio, Kokuoz, James et al., 2008); modern bioceramic osteoconductive materials (Kusyak, Petranovska, Dubok et al., 2021; Hench and Fielder, 2004; Dutra, Pereira, Serakides et al., 2008); highly efficient biocompatible nanosized phosphors sensitive to biologically safe high-permeability "soft" X-ray radiation (Jing, Guo, Diao et al., 2015; Patro, Bharathi, and Chandra, 2014; Tang, Hu, Elmenoufy et al., 2015; Kasturi, Marikumar, and Vaidyanathan, 2018), and photosensitizers with specified spectral characteristics of luminescence and absorption, respectively.

Based on the focus of the work on the synthesis and study of the properties of nanodispersed phosphors for PDT of tumor diseases in cranial organs and bone tissues, it may be promising for research to use X-ray luminescent nanosized particles (NPs) LaF_3 (Liu, Chen, Wang et al., 2008; Hsiu-Wen, Chien-Hao, Chien-Hsin et al., 2020) and $LaPO_4$ (Medkov, Steblevskaya, and Belobeletskaya, 2017), activated by rare earth ions, obtained by precipitation from aqueous solutions. Their research and optimization of technology of synthesis of nanostructures, comparison of results and study of luminescence properties are expedient to carry out with use of ultraviolet (UV) radiation with which work is much simpler, than with X-ray radiation.

Therefore, the aim of this work is the synthesis of nanodispersed phosphors based on lanthanum fluoride and phosphate activated by terbium ($LaF_3:Tb^{3+}$ and $LaPO_4:Tb^{3+}$, respectively), promising for use in photodynamic therapy and optopharmacology, study of their structural properties and luminescence as well as the possibility of their use in nanocomposites with magnetically sensitive nanosized Fe_3O_4 carriers and 60S bioactive glass.

Materials and Methods of Research

Synthesis of nanodispersed crystals LaF_3 and $LaF_3:Tb^{3+}$ was carried out according to the method (Liu, Chen, Wang et al., 2008; Hsiu-Wen, Chien-Hao, Chien-Hsin et al., 2020) by coprecipitation of components from aqueous and alcoholic (methanol) solution. As precursors were used: $La(NO_3)_3 \cdot 6H_2O$, $TbCl_3$, NH_4F, $(NH_4)_2HPO_4$. All reagents used were "chemically pure". Distilled water and methanol were used as solvents. The use of water medium for the synthesis has a number of advantages because it is environmentally safely and compatible with methods for obtaining of NCs based on magnetite NPs and magnetic fluids, as well as bioactive glass based on phosphates and silicates. It is known that LaF_3 is insoluble in water and can form crystal hydrates of $LaF_3 \cdot 0.5H_2O$.

For the synthesis of $LaF_3:Tb^{3+}$ it was taken: 18.473 g of $La(NO_3)_3 \cdot 6H_2O$ (42.664 mmol), 5.708 g of $TbCl_3$ (10.769 mmol), 4.773 g of NH_4F (127.992 mmol). The ratio of reagents provided the composition of the synthesized samples, which corresponds to the formula $La_xTb_{1-x}F_3$, $x = 0.8$.

The following synthesis variants have been developed to obtain $LaF_3:Tb^{3+}$ samples.

1. La^{3+} and Tb^{3+} salts in molar ratios of 4:1 were successively dissolved in a minimum volume of distilled water (or methanol). With constant stirring, the solution was added dropwise with the contents of F^- ions in the ratio La:F as 1:3. The reaction solution was stirred at room temperature for 2 hours. The product was centrifuged, washed three times with deionized water and dried at room temperature.
2. In the same ratios and succession, the chosen components were introduced into a reactor, and synthesis was carried out at 75°C. In this case, an increase in temperature leads to a relatively small increase in the size of the primary particles (~ 10-15%), a decrease in the degree of aggregation and the formation of a more ordered crystal structure.
3. The obtained solution containing $LaF_3:Tb^{3+}$ NPs was transferred into an autoclave with programmed heating/cooling at a rate of $1°C \cdot min^{-1}$

and kept at 150°C for 24 hours. The obtained products were separated, washed and dried at 60°C. It is known that the treatment with elevated temperature and pressure leads to disappearance of small crystals, evolution of crystal shape to hexagonal, and formation of porous surface of massive samples.

The use of phosphates in the research of this work is mainly due to their better biocompatibility and biosafety when interacting with bone tissue.

To obtain samples of nanodispersed crystals of $LaPO_4:Tb^{3+}$, salts of La^{3+} and Tb^{3+} in molar ratios of 3:1.1 were successively dissolved in 100 mL of deionized water. It was taken per 100 mL of a solution: 0.13 g of $La(NO_3)_3 \cdot 6H_2O$ (3.0 mmol), 0.041 g of $TbCl_3$ (1.1 mmol), 0.049 g of $NH_4H_2PO_4$ (4.1 mmol). To the mixture of salts, with constant stirring, a solution of PO_4^{3-} ions was added dropwise for 2 hours, in the ratio $La^{3+}:PO_4^{3-}$ as 3:4.1. A highly dispersed precipitate of white color is formed. The product was centrifuged, washed three times with deionized water and dried at room temperature. The used ratio of reagents provided the composition of the synthesized samples, which corresponds to the formula $La_xTb_{1-x}PO_4$, $x = 0.8$.

The average size of lanthanum fluoride and phosphate crystals depended on the conditions of their synthesis.

The synthesis of nanosized magnetite in a single-domain state was performed according to the method (Gorobets', Gorobets', Gorbyk et al., 2018).

Research of the samples was performed by differential thermal analysis (DTA) using a derivatograph Q-1500D of MOM company (Hungary).

In this work, $Fe_3O_4/LaF_3:Tb^{3+}$ NC (Kusyak, Petranovska, Turanska et al., 2021; Mangaiyarkarasi, Chinnathambi, Karthikeyan et al., 2016; He, Xie, Ding et al., 2009) was synthesized as follows: first to starting freshly synthesized magnetite, washed with distilled water to pH = 7, a solution of salts of La^{3+} and Tb^{3+} was added in the ratio of the quantity of active hydroxyl groups on its surface to the number of La^{3+} ions as 1:1 (according to DTA, the concentration of hydroxyl groups on the surface of Fe_3O_4 was 2.2 mmol·g^{-1}) and left for 24 hours for adsorption saturation of the surface. Then the salt solution was drained, Fe_3O_4 particles with adsorbed La^{3+} and Tb^{3+} were washed three times and poured with 50 mL of deionized water. After that, with constant stirring, the solution was added dropwise with the contents of F^- ions in the ratio La:F as 1:3 (it is assumed that all ions La^{3+} and Tb^{3+} are adsorbed on the surface of Fe_3O_4). The obtained $Fe_3O_4/LaF_3:Tb^{3+}$ NC was washed with distilled water and dried at 60°C. It should be noted that the chosen method of synthesis of $Fe_3O_4/LaF_3:Tb^{3+}$ NC promotes the formation of their structure by core–shell type (Abramov, Turanska, and Gorbyk, 2018a; Abramov, Turanska, and Gorbyk, 2018b) (core Fe_3O_4 – shell $LaF_3:Tb^{3+}$).

The synthesis of sol-gel bioglass (BG 60S) was carried out according to the method (Kusyak, Petranovska, Dubok et al., 2021). 60S glass has a composition (mol %): 60% SiO_2, 36% CaO, 4% P_2O_5. The synthesis was carried out by sol-gel method using such precursors: tetraethyl orthosilicate (TEOS) $(C_2H_5O)_4Si$, triethylphosphate (TEP) $(C_2H_5O)_3PO$, ethanol C_2H_5OH, calcium nitrate tetrahydrate $(Ca(NO_3)_2 \cdot 4H_2O)$, lanthanum(III) nitrate hexahydrate $(La(NO_3)_3 \cdot 6H_2O)$, 59% solution of nitric acid (HNO_3) (all reagents of qualification "chemically pure" (Merck Schuchardtohg, Germany)). For the synthesis of 60S glass, mass ratios of precursors were: $(C_2H_5O)_4Si:(C_2H_5O)_3PO:(Ca(NO_3)_2 \cdot 4H_2O):H_2O:C_2H_5OH = 8.59:1:5.85:9:3$. To obtain sol-gel glass, TEOS, TEP and ethanol are first poured in the above proportions, mixed on a magnetic stirrer for 30 min, then sonication is applied for 5 min. For hydration and obtaining of sol, nitric acid is added, the mixture is stirred again with a magnet for 30 min and again sonicated for 5 min. Separately an aqueous solution of calcium nitrate is prepared, mixing the specified quantities of components on a magnetic stirrer for at least 10 min. Then a solution of calcium nitrate is added to a sol, mixed on a magnetic stirrer for at least 40 min, sonicated for 5 min. To complete the polycondensation processes, a sol was left for 24 h at room temperature, then it was heated in a sealed container in a dry oven for 24 h at 60°C. The resulting gel is kept at least 48 h at 120°C and then heated slowly (at least 4 h) to 700-900°C and calcined at this temperature for 2 h.

To obtain X-ray luminescent sol-gel glass, the method (Kusyak, Petranovska, Dubok et al., 2021) was modified: after the process of hydration of TEOS and TEP, with constant stirring, previously synthesized X-ray phosphor ($LaF_3:Tb^{3+}$, $LaPO_4:Tb^{3+}$) was added and sonicated for 5 minutes. All other stages of the synthesis were carried out similarly to (Kusyak, Petranovska, Dubok et al., 2021). The amount of X-ray phosphor was ~ 1.5% (wt.) of the prepared X-ray luminescent sol-gel glass.

Structural studies of the obtained samples were performed by powder X-ray diffraction method (XRD) using DRON-UM1 diffractometer with Fe filtered Cu K_α radiation, the Bragg-Brentano focusing, in 2θ range of 10-80° with a step of 0.05°, exposure of 1 s. Phase identification was performed on the basis of PDF-2 database. The average crystal size was determined by the width of the corresponding most intense line according to the Scherrer equation.

Biocompatibility test of the synthesized model samples was realized on human A549 non-small cellular lung cancer cell line. The cell line was provided by the bank of cells from human and animal tissues of R.E. Kavetsky Institute of Experimental Pathology, Oncology and Radiobiology of NAS of Ukraine. A549 cells were grown in RPMI medium (Biowest, France) supplemented with 1% fetal bovine serum (FBS, Biowest, France), 1% penicillin-streptomycin and 1% L-glutamine (a complete medium) in the incubator at 37°C and atmosphere of 5% CO_2. For the biocompatibility experiments, a total of $3 \cdot 10^4$ A549 cells per well were seeded in 24-well plates and cultivated for 24 h. After this time, the medium in corresponding wells was supplemented with 0.1 mg·mL^{-1} of $LaF_3:Tb^{3+}$ nanoparticles preliminary diluted in RPMI medium, whereas wells free of nanoparticles served as a control. Cells in the respective wells were continued to cultivate under the same conditions for 72 hours. After 72 hours of cultivation, the number of cells was counted in the Goryaev chamber. To determine the effect of $LaF_3:Tb^{3+}$ nanoparticles on cell survival, the average number of cells in all control wells was determined. This number was used to normalize cell count in each well with respective concentration of $LaF_3:Tb^{3+}$. Biocompatibility (in percent) was expressed as mean cell count in the well in the presence of $LaF_3:Tb^{3+}$ relative to control ± standard deviation (mean ± SD) times 100 with the "n" – number of wells (Foster, Oster, Mayer et al., 1998). Cell survival assay was realized in four independent experiments.

Magnetic hysteresis loops of ensembles of Fe_3O_4 NPs were measured using a laboratory vibration magnetometer of Foner type at room temperature (Abramov, Turanska, and Gorbyk, 2018a; Abramov, Turanska, and Gorbyk, 2018b). To prevent interaction, demagnetized nanoparticles were distributed in a paraffin matrix with a volume concentration of ~ 0.05. For comparison, we used materials with a known value of the specific saturation magnetization (σ_s): tested specimen of nickel and Fe_3O_4 NPs (98%) manufactured by Nanostructured & Amorphous Materials Inc., USA. In relation to the reference sample, the measurement error σ_s did not exceed 2.5%.

To study morphology and size distribution of NPs, we used their colloidal solutions in water. The size and shape of NPs were determined by electron microscopy using a transmission electron microscope JEOL 1200 EX (Tokyo, Japan) at an acceleration voltage of 120 kV. The samples were diluted in deionized water, placed onto carbon coated copper grid (EM Resolutions Ltd), and dried at room temperature for 12 hours.

Samples were investigated by scanning electron microscopy using a device JSM-6060 LA (JEOL, Japan).

Particle size distributions in water suspension were plotted based on photon correlation spectroscopy (PCS) using an analyzer Nanophox (Sympatec, Germany).

Specific surface area and pore size distribution for nanodispersed samples were determined by thermo-stimulated low-temperature nitrogen desorption using NOVA 1200e Surface Area & Pore Size Analyzer (Quantachrome, USA). The size of NPs was estimated by the formula $D_{BET} = 6/(\rho S_{BET})$, where ρ is the particle density, S_{BET} is the value of the specific surface area calculated by Brunauer, Emmett and Teller (BET) theory of polymolecular adsorption. The value of S_{BET} was estimated by measuring the isotherms of physical adsorption – desorption of nitrogen at -196°C in the range of relative pressures $P/P_0 = 0.05$-0.20.

Infrared spectra of absorption were recorded on FT-IR spectrometer Tensor 27 (Bruker Optik GmbH, Germany) in the range of 4000-400 cm^{-1}, using KBr granules, with a resolution of 2 cm^{-1}.

To indicate the stability of particle suspensions and pH values of isoionic point (pH_{IIP}), zeta potential of particles was measured with a laser Doppler electrophoresis (LDE) instrument (Nano Series, Malvern Instrument Ltd., UK) with 2 g·L^{-1} sample concentration in 0.15 mol·L^{-1} NaCl (NSS). The potential was determined three times, the mean values and standard deviations were calculated.

Study of acid-base surface characteristics and potentiometric measurements of LaF$_3$ and LaF$_3$:Tb^{3+} suspensions were carried out using a device I-160M. The acid-base properties of the surface of the samples were investigated by the method of pH-metry of individual samples, which allows to estimate the integral acidity of the surface in the study of changes in pH of the aqueous suspension of samples. 0.01 g of the test sample was added into conical flasks, then 5 mL of electrolyte solution (NSS) of different pH was added (pH = 2.5-12 is set by adding 0.01 M solutions of NaOH or HCl). Suspensions in closed conical flasks were mixed on a shaker for 2 hours. A solution was separated from a sample by centrifugation, and equilibrium pH (pH_{eq}) was measured. The difference in the acidity values of the solutions before (pH$_0$) and after (pH$_{eq}$) shows a change ($\pm\Delta pH$) as a result of hydrolytic adsorption $\pm\Delta pH = pH_0 - pH_{eq}$.

Analyzing the dependences of hydrolytic adsorption, the conclusion is made about the acid-base nature of surface centers of LaF$_3$ and LaF$_3$:Tb^{3+}. The change in the pH of the aqueous suspension occurs due to the adsorption processes involving H$^+$ and OH$^-$ ions, and the dissociation of water molecules by heterolytic mechanism. Determination of integrated indicators of acid-base properties of suspensions is based on the calculation of ionization constants of surface centers K_1 and K_2, which characterize the following surface equilibria:

$$-E_{(S)}\cdots H^+ \square \xrightarrow{K_1} -E_{(S)}^0 + H^+ \qquad (1)$$

$$-E_{(S)}\cdots H-OH \square \xrightarrow{K_2} -E_{(S)}\cdots OH^- + H^+ \qquad (2)$$

$-E_{(S)}\cdots H^+$ – surface group that arises on the surface due to adsorption of H⁺ from the liquid phase;

$-E_{(S)}^0$ – neutral group of the surface;

$-E_{(S)}\cdots OH^-$ – surface group that arises on the surface due to dissociation of H–OH and transition of H⁺ to the liquid phase;

H^+ – hydrogen ions in the suspension phase;
K_1 and K_2 – ionization constants.

Using (1, 2), the value of ΔpH_i is calculated. Using the definition for K_1 and K_2, the expression is derived:

$$K_1 = \frac{[H^+]\cdot[-E_{(S)}^0]}{[-E_{(S)}\cdots H^+]} \qquad (3)$$

$$K_2 = \frac{[H^+]\cdot[-E_{(S)}\cdots OH^-]}{[-E_{(S)}\cdots H-OH]} \qquad (4)$$

The change in [H⁺] in the suspension is determined by the equation

$$\Delta[H^+] = [H_0^+] + [H_{eq}^+], \qquad (5)$$

$[H_{eq}^+]$ – equilibrium concentration of hydrogen ions in suspension, $[H^+_0]$ – the concentration of hydrogen ions in the initial solution; $\Delta[H^+]$ – the corresponding difference in ion concentrations:

$$\Delta[H^+] = [H^+]_1 + [H^+]_2 = ([-E \cdots H^+]_0 - [-E \cdots H^+]) + ([-E \cdots OH^-] - [-E \cdots OH^-]_0) \quad (6)$$

$\Delta[H^+]_1$ – change in H⁺ concentration associated with surface groups $-E_{(S)} \cdots H^+$;

$\Delta[H^+]_2$ – change in H⁺ concentration associated with surface groups $-E_{(S)} \cdots OH^-$;

$[-E \cdots H^+]_0$ and $[-E \cdots OH^-]_0$ – concentrations of protonated and deprotonated surface groups, respectively, which characterize the initial state of the surface.

Taking into account the use of NaCl solution as an indifferent electrolyte, the values of ionic strength and activity coefficients were determined used for calculations. Thus, using equation (6) and the values of $\Delta[H^+]_1$ and $\Delta[H^+]_2$ for each of the pH values, calculated from experimental data, concentrations of protonated and deprotonated surface groups $[-E \cdots H^+]$ and $[-E \cdots OH^-]$, respectively, are determined.

The total concentration of surface groups $C_{(S)}$ is defined as:

$$C_{(S)} = [-E^0] + [-E \cdots H^+] + [-E \cdots OH^-] \quad (7)$$

Equations (3, 4, 7) are used to determine the parts of negatively α^-, positively α^+ charged and neutral α^0 groups in the studied pH range:

$$\alpha^- = \frac{K_1 \cdot K_2}{[H^+] \cdot K_1 + K_1 \cdot K_2 + [H^+]^2} \quad (8)$$

$$\alpha^+ = \frac{\left[H^+\right]^2}{\left[H^+\right] \cdot K_1 + K_1 \cdot K_2 + \left[H^+\right]^2} \quad (9)$$

$$\alpha^0 = \frac{\left[H^+\right] \cdot K_1}{\left[H^+\right] \cdot K_1 + K_1 \cdot K_2 + \left[H^+\right]^2} \quad (10)$$

Based on the results of calculations, a diagram was constructed for the dependence of concentration of active centers of surface of the particles on pH of the medium. The results of three parallel measurements were processed by the methods of mathematical statistics, the calculated error of the accuracy of direct measurement does not exceed 2.5%. Estimation of the error of measurement results was performed taking into account the values of accuracy of measuring instruments.

Excitation of the luminescence of the samples was carried out by UV beams using a lamp DRSH-500, radiation passed through a UV filter MidOpt BP324.

Measurement of X-ray excited luminescence spectra was realized by a spectrometer Ocean Optics USB2000 and OmniDriverCSharpDemo software with parameters: integration time of 10 s, averaging 30 times. Spectra were measured by the optical fiber FC-UV-400-2 brought into a diffractometer, with a collimator nozzle. Distance from particles to a collimator was approximately 30 mm. Analysis of luminescence spectra was performed by fitting certain areas of the spectra in accordance to the measured peaks by the Gaussian curve.

For excitation of particles, we used X-ray source of CuKα type (λ = 0.154056 nm) from a powder diffractometer ARXD of Proto company. Irradiation of all the studied particles was realized under such working parameters of X-ray source: U = 30 kV, I = 20 mA. A beam of X-rays fell on a holder with test particles at the angle of 40° during all the time of measuring of spectra (~ 300 s).

Results and Discussion

It is known that lanthanum(III) fluoride forms colorless crystals of hexagonal syngony, spatial group P3c1, elementary lattice parameters $a = 0.7185$ nm, $c = 0.7351$ nm, $Z = 6$ (ICDD: 78-1864). Figure 1, *a, b, c, d* shows the results of X-ray diffraction of the synthesized (variant 1) samples of lanthanum fluoride activated by terbium. It's seen that the diffraction patterns of $LaF_3:Tb^{3+}$ samples synthesized in water (Figure 1, *a*) and methanol (anhydrous medium) are not fundamentally different (Figure 1, *b*).

Under the experimental conditions, fairly perfect $LaF_3:Tb^{3+}$ crystals were formed during crystallization at a temperature of 75 °C (synthesis variant 2). Figure 1, *c* shows the diffraction patterns of samples LaF_3 (curve 1) and $LaF_3:Tb^{3+}$ (curve 2), synthesized according to variant 2. Preliminary values of the parameters a and c of the hexagonal elementary lattice of LaF_3 were determined by lines (300) and (302) using a quadratic form for hexagonal syngony, which connects the interplanar distance, Miller indices and lattice parameters.

The values of the parameters specified by the method of least squares are given in Table 1. The values of the volume of the elementary lattice in the crystal structure of the studied samples are also given. It's seen that the presence of Tb^{3+} leads to a decrease in the lattice parameters of hexagonal $LaF_3:Tb^{3+}$. The average size of LaF_3 crystals was ~ 15 nm, and $LaF_3:Tb^{3+}$ ~ 14 nm (Figure 1, *c*, synthesis variant 2). Taking into account the sizes of La^{3+} and Tb^{3+} ions, the assumption was made about the possibility of replacing of La^{3+} ions with Tb^{3+} ions in samples $LaF_3:Tb^{3+}$ (Hsiu-Wen, Chien-Hao, Chien-Hsin et al., 2020; Mangaiyarkarasi, Chinnathambi, Karthikeyan et al., 2016; Zhang, Braun, Pallaoro et al., 2011).

$LaF_3:Tb^{3+}$ crystals synthesized in an autoclave (variant 3) showed the most perfect structure (Figure 1, *d*), with an average size of ~ 20 nm.

Table 1. Parameters of the elementary lattice in the crystal structure of the studied samples

Specimen	a, Å	c, Å	The volume of the crystal lattice, Å^3
LaF_3 ICDD: 78-1864	7.185	7.351	328.65
LaF_3	7.186	7.353	328.8
$LaF_3:Tb^{3+}$	7.144	7.318	323.4

Figure 1, *e* shows the diffraction patterns of typical samples of LaPO$_4$:Tb^{3+}. According to X-ray diffraction data (ICDD 46-1439), the synthesized sample of lanthanum phosphate LaPO$_4$:Tb^{3+} consists of a phase of LaPO$_4$·0.5H$_2$O. The average crystal size was ~ 10 nm.

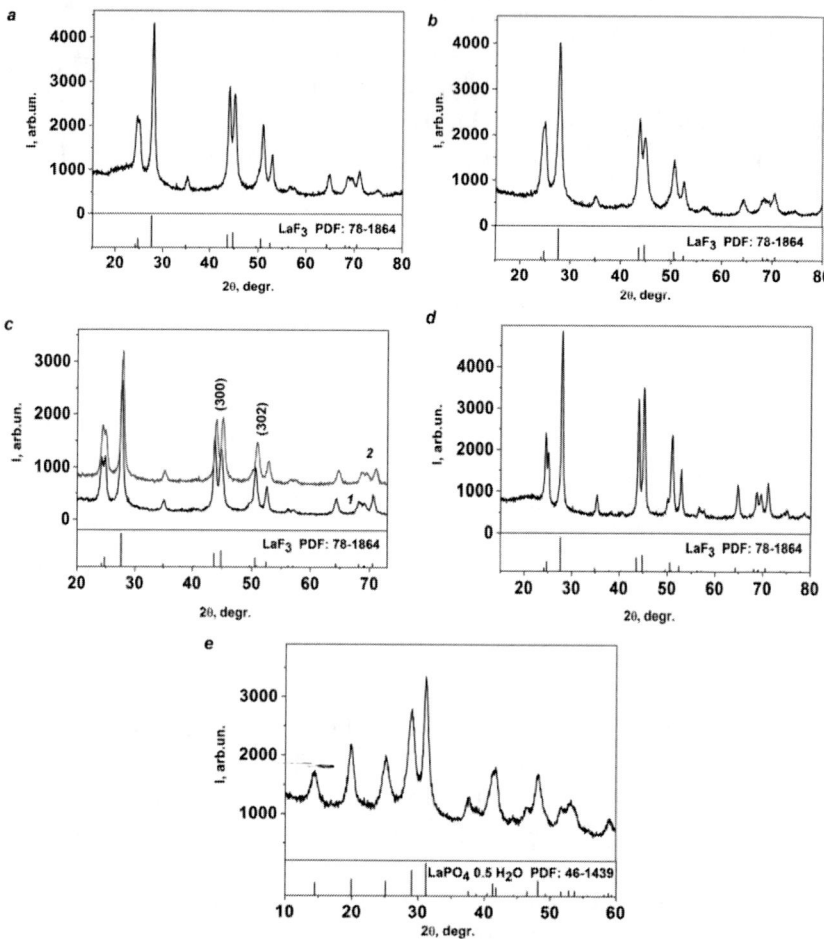

Figure 1. XRD patterns of samples: LaF$_3$:Tb^{3+}, synthesized according to variant 1 in water (*a*), methanol (*b*); LaF$_3$ (1) and LaF$_3$:Tb^{3+} (2), according to variant 2 (*c*); in autoclave, variant 3 (*d*); LaPO$_4$:Tb^{3+} (*e*).

Figures 2, 3 show TEM images of synthesized (variant 2) nanocrystals LaF_3 (Figure 2) and $LaF_3:Tb^{3+}$ (Figure 3), as well as their size distributions according to TEM (*b*) and photon correlation spectroscopy (PCS) (*c*). Characteristically, as measured by independent methods, in the ensemble of $LaF_3:Tb^{3+}$, the average nanoparticle size is significantly smaller than the average nanoparticle size in the ensemble of LaF_3. These data correlate with the results of determination of the average particle size by the Scherrer formula.

Figure 4 shows TEM (*a*) and SEM (*b*) images of synthesized nanocrystalline samples of $LaPO_4:Tb^{3+}$, size scale of 20 nm (*a*) and 1 μm (*b*). It's seen that the synthesized nanocrystals of fluoride and phosphate are prone to aggregation and formation of chain structures.

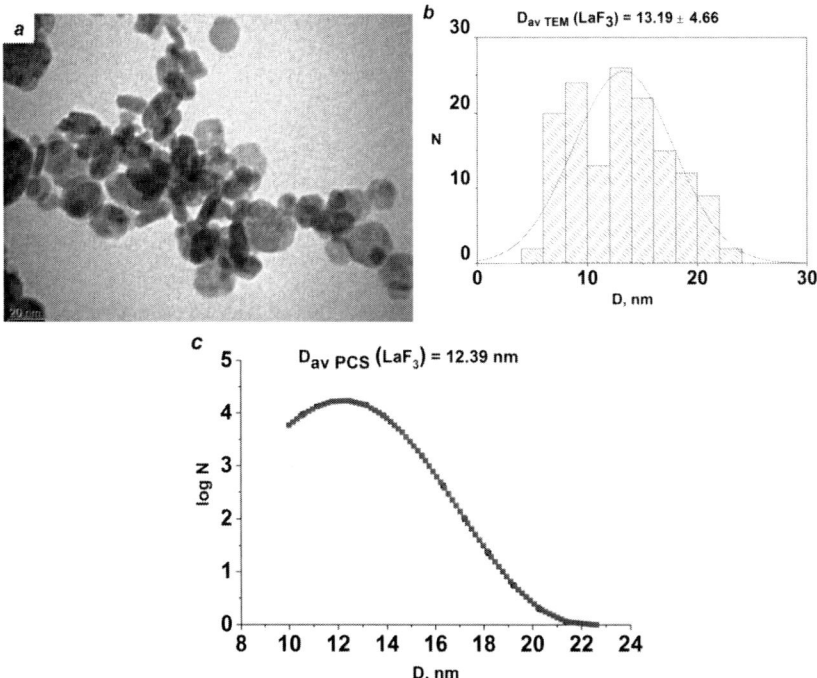

Figure 2. TEM images of LaF_3 NPs: scale of 20 nm (*a*); the corresponding particle size (*D*) distribution according to TEM (*b*) and PCS (*c*). Average size of particles, nm: 13.19 ± 4.66 (*b*) and 12.39 (*c*) nm. Synthesis variant 2. *N* – number of particles.

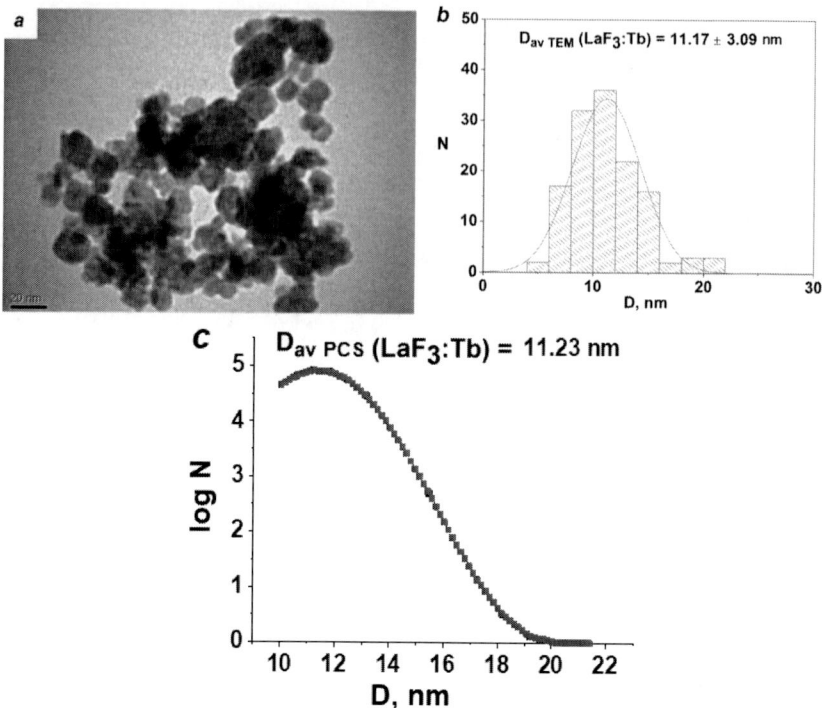

Figure 3. TEM images of LaF$_3$:Tb^{3+} NPs: scale of 20 nm (*a*); the corresponding particle size (*D*) distribution according to TEM (*b*) and PCS (*c*). Average size of particles, nm: 11.17 ± 3.09 nm (*b*) and 11.23 (*c*) nm. Synthesis variant 2. *N* – number of particles.

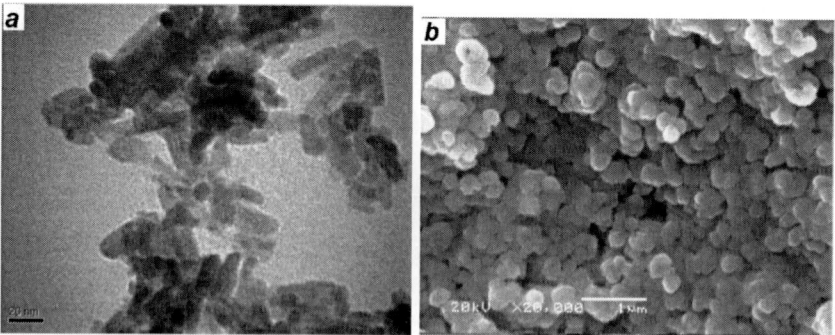

Figure 4. TEM (a) and SEM (b) images of typical LaPO$_4$:Tb^{3+} samples: size scale of 20 nm (a); 1 µm (b).

The results of studies of the influence of LaF$_3$:Tb^{3+} NPs on the survival of human A549 non-small cellular lung cancer cells are shown in Figure 5.

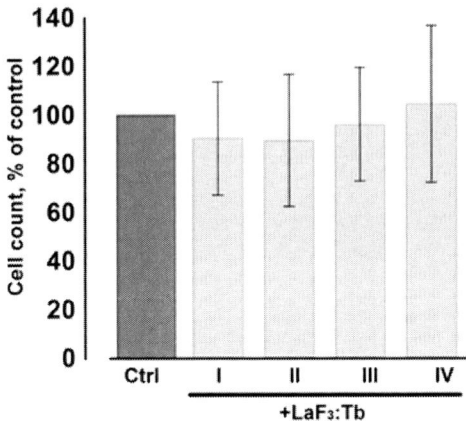

Figure 5. Influence of LaF$_3$:Tb^{3+} NPs on the survival of human A549 non-small cellular lung cancer cells.

Changes in the number of cells cultivated in the presence of 0.1 mg·mL^{-1} LaF$_3$:Tb^{3+} (color bars, mean ± standard deviation) compared to control (Ctrl, light gray bar, 100%) are shown in the diagram ($n = 8$ for each bar). Roman numerals denote four independent experiments. These data confirm the high biocompatibility of LaF$_3$:Tb^{3+} nanoparticles with respect to human A549 non-small cellular lung cancer cell line.

Low-temperature isotherms of adsorption/desorption of nitrogen (1) and pore size distribution curves (2) of nanodispersed LaF$_3$ (*a*) and LaF$_3$:Tb^{3+} (*b*), respectively, are given in Figure 6. According to research, there are some differences in these characteristics of the synthesized samples. Thus, LaF$_3$ samples are characterized by a BET specific surface area (S_{BET}) of 34.73 m^2·g^{-1}, and LaF$_3$:Tb^{3+} – 68.59 m^2·g^{-1}, whereas the total pore volume increases from 0.205 to 0.248 cm^3·g^{-1}, and the average pore diameter decreases from 32.88 nm to 16.75 nm. An increase in the values of specific surface area of the samples indicates a decrease in the average particle size, and the nature of the isotherms, as well as the presence of a hysteresis loop – the presence of pores, probably represented by interparticle space.

In FT-IR absorption spectra (Figure 7), the bands of deformation vibrations (3420 and 1630 cm^{-1}, asymmetric (v_{as}) and symmetric (v_s), respectively) are observed attributed to OH-bonds of water molecules forming a crystal hydrate with lanthanum fluoride, as well as adsorbed on the surface of particles.

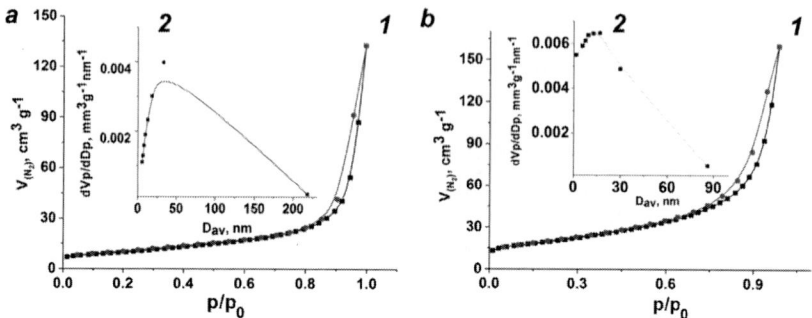

Figure 6. Low-temperature nitrogen adsorption/desorption isotherms (1) and pore size distribution curves (2) for samples LaF_3 (*a*) and $LaF_3:Tb^{3+}$ (*b*). $V(N_2)$ – volume of adsorbed nitrogen, V_p – pore volume, D_p – pore diameter, p_0 – initial gas pressure, p – nitrogen pressure after adsorption.

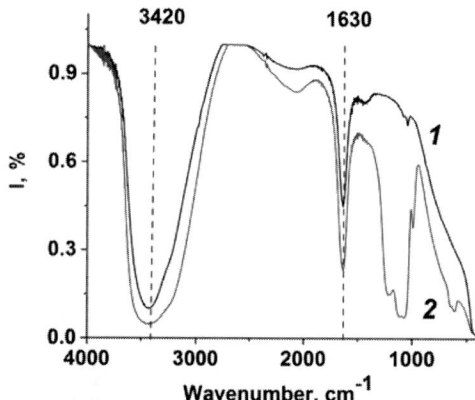

Figure 7. FT-IR spectra of LaF_3 (1), $LaF_3:Tb^{3+}$ (2).

Figures 8, *a, b* show the results of ζ-potential measurements for nanodispersed samples LaF_3 and $LaF_3:Tb^{3+}$ in NSS medium as a function of pH values. For LaF_3 NPs in a colloidal system (Figure 8, *a*), the values of zeta potential in the range of 40.32 ± 3.32-30.87 ± 4.56 mV indicate a sufficient stability of dispersions in the range of pH ~ 2-6. The high surface charge density and fairly high electrostatic repulsion between NPs in colloidal systems provide their stability. At pH ~ 7 NPs do not carry a charge, corresponding to their isoelectric point. At pH 7-8, the colloidal stability is lost, and LaF$_3$ NPs are able to agglomerate.

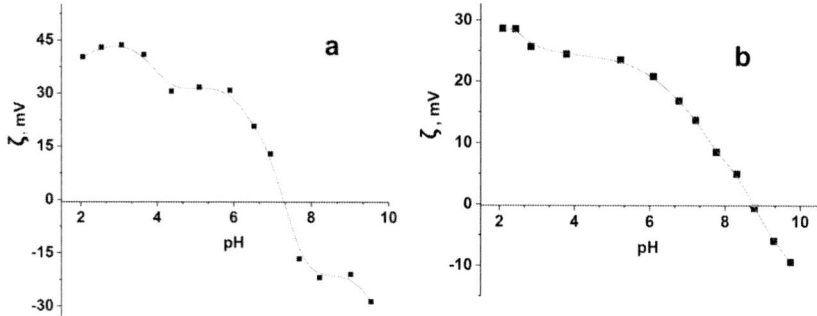

Figure 8. pH dependences of the ζ-potentials of suspensions of NPs in NSS medium: LaF$_3$ (*a*), LaF$_3$:Tb^{3+} (*b*).

For LaF$_3$:Tb^{3+} NPs in the experimental range of pH ~ 2-6, analysis of zeta potential (Figure 8, *b*) showed slightly lower values from 28.67 ± 1.53 to 20.8 ± 0.7 mV, which explains the lower stability of suspensions in NSS medium. In addition, it should be noted that the isoionic point for LaF$_3$:Tb^{3+} NPs is shifted to a pH value of ~ 8.5.

Table 2. pH dependence of the concentration of protonated $^{-E_{(S)}\cdots H^+}$ and deprotonated $^{-E_{(S)}\cdots OH^-}$ active centers of surface of LaF$_3$

pH	$[-E_{(S)}\cdots H^+] \cdot 10^{-3} \pm X \cdot 10^{-4}$ mol g^{-1}	pH	$[-E_{(S)}\cdots OH^-] \cdot 10^{-4} \pm X \cdot 10^{-7}$ mol g^{-1}
2.43	87.50 ± 43.01	5.82	3.805 ± 183.92
3.43	12.05 ± 4.05	8.06	1.285 ± 47.35
4.3	3.31 ± 1.41	9.45	0.578 ± 25.31
5.24	0.027 ± 0.023	10.96	0.0037 ± 0.0031
		11.96	2.98 ·10^{-7} ± 1.32·10^{-5}
		12.78	2.131·10^{-8} ± 8.44·10^{-7}
$\sum [-E_{(S)}\cdots H^+] \cdot 10^{-3} \pm X \cdot 10^{-4}$ mol g^{-1}		$\sum [-E_{(S)}\cdots OH^-] \cdot 10^{-4} \pm X \cdot 10^{-6}$ mol g^{-1}	
102.9 ± 12.128		5.66 ± 4.27	

For NSS medium, the values of $[H^+]_0$ and $[H^+]_{eq}$ were experimentally obtained, which were used for calculation of the changes in activity Δa_H and hydrogen ion concentration $\Delta[H^+]$, and the concentrations of protonated and deprotonated active centers of surface of LaF_3 and $LaF_3:Tb^{3+}$ were found at different pH (6), as well as the value of the total concentration of active centers (7) (shown in Tables 2, 3).

Table 3. pH dependence of the concentration of protonated $-E_{(S)} \cdots H^+$ and deprotonated $-E_{(S)} \cdots OH^-$ active centers of surface of $LaF_3:Tb^{3+}$

pH	$[-E_{(S)} \cdots H^+] \cdot 10^{-3} \pm X \cdot 10^{-4}$ mol g^{-1}	pH	$[-E_{(S)} \cdots OH^-] \cdot 10^{-4} \pm X \cdot 10^{-6}$ mol·g^{-1}
2.43	58.21 ± 22.6	6.01	2.71 ± 27.66
3.51	9.5 ± 3.11	7.5	1.79 ± 6.19
4.24	3.35 ± 5.36	9.2	2.14 ± 45.61
$\Sigma [-E_{(S)} \cdots H^+] \cdot 10^{-3} \pm X \cdot 10^{-4}$ mol·g^{-1}		10.94	0.043 ± 0.45
71.07 ± 0.01		11.96	1.37 ·10^{-6} ± 0.38
		12.73	1.40 ·10^{-8} ± 4.51·10^{-5}
$[-E_{(S)} \cdots OH^-] \cdot 10^{-5} \pm X \cdot 10^{-6}$ mol·g^{-1}		$\Sigma [-E_{(S)} \cdots OH^-]$ 10^{-4} ± X·10^{-6} mol·g^{-1}	
5.41	8.283 ± 1.23	7.52 ± 13.51	

Using expressions (3), (4), the values of the ionization constants of surface centers K_1 and K_2 and the corresponding values of pK_1 and pK_2 of active centers of surface of LaF_3 and $LaF_3:Tb^{3+}$ nanostructures were calculated for different pH in NSS medium (Table 4).

Table 4. pH dependence of pK values for active centers of surface of LaF_3 and $LaF_3:Tb^{3+}$ nanostructures

$[-E_{(S)} \cdots H^+]$				$[-E_{(S)} \cdots OH^-]$			
pK_1				pK_2			
pH	LaF$_3$	pH	LaF$_3$:Tb^{3+}	pH	LaF$_3$	pH	LaF$_3$:Tb^{3+}
2.43	0.44±0.05	2.43	0.38±0.01	5.82	9.7±0.41	6.01	9.85±0.12
3.43	0.63±0.05	3.51	0.76±0.006	8.06	10.79±0.56	7.5	10.33±0.23
4.3	1.21±0.12	4.24	1.23±0.08	9.45	11.49±0.38	9.2	10.19±0.15
5.24	-0.3±0.24			10.96	17.84±0.11	10.94	13.57±0.26
				11.96	23.37±0.15	11.96	22.31±0.17
				12.78	25.39±0.49	12.73	25.38±0.39

Using expressions (3), (4), (7) and (8-10), the parts of negatively α^-, positively charged α^+ and neutral α^0 active centers of LaF$_3$ and LaF$_3$:Tb^{3+} nanostructures in physiological saline were calculated in the range of pH 2.4-12.7 (Figure 9, a, b).

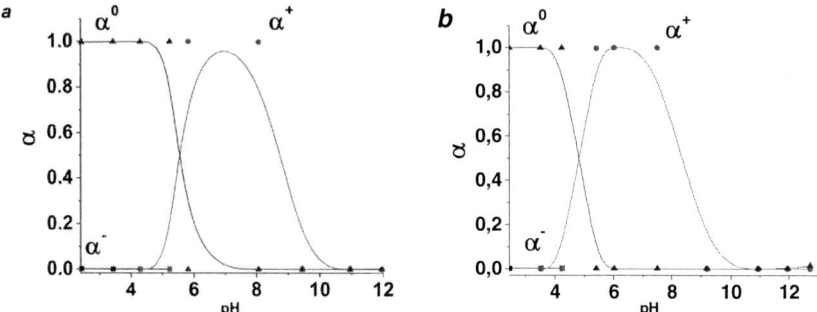

Figure 9. pH dependence of the part of neutral α^0, negatively α^- and positively α^+ charged active surface centers of nanostructures LaF$_3$ (a) and LaF$_3$:Tb^{3+} (b).

According to the results of studies, in the pH range of 2.4-5.3, neutral (α^0) centers are dominant for the surfaces of LaF$_3$ and LaF$_3$:Tb^{3+} (Figure 9, a, b). In the pH range of 2.4-5.3, the part α^- of strong base centers capable of protonation, for which pK = 0.44-1.23 (Table 4), is very small (87.5·10^{-3}-0.027·10^{-3} mol g^{-1} for LaF$_3$; 58.21·10^{-3}-3.35·10^{-3} mol g^{-1} for LaF$_3$:Tb^{3+} (Tables 2, 3)). In the pH range of 6-9.5, the centers α^+ able to deprotonation are dominant, with weak acidic properties (pK 9.7-11.49 (Table 4)), and their concentration is low (3.8-0.57·10^{-4} mol g^{-1} for LaF$_3$, 2.71-2.14·10^{-4} mol g^{-1} for LaF$_3$:Tb^{3+} (Tables 2, 3)). Thus, in the pH range of 2.4-12.7, the detected active centers of surface can be represented as acidic (La^{3+}) and base (F$^-$) Lewis centers, as well as base Bronsted centers (OH$^-$ groups) (Kemnitz and Coman, 2016; Tressaud, 2010). It should be noted that at pH values above 10, the hydrolytic activity of surface centers is strongly inhibited. The obtained data are useful for optimization of conditions of adsorption immobilization of molecules of photosensitive substances (sensitizers) on the surface of phosphors based on lanthanum fluoride.

Figure 10, a shows photographs of photoluminescence of samples of colloidal systems based on water and LaF$_3$ when excited by UV radiation. It's seen that in LaF$_3$ samples non-activated by ions of rare earth elements, luminescence is almost absent. Effective "green" photoluminescence is observed in nanodispersed samples of LaF$_3$:Tb^{3+} (Figure 10, a). With an increase in the concentration of $LaF_3:Tb^{3+}$, the luminescence intensity increases. The luminescence spectrum can be effectively controlled by changing the rare earth admixtures in the structure of LaF_3 matrix: for example, the use of Eu^{3+} allows to obtain $LaF_3:Eu^{3+}$ nanocrystals with intense red photoluminescence under UV excitation (Figure 10, b).

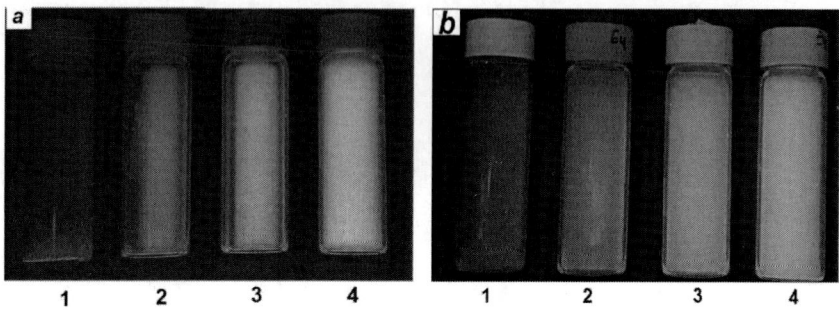

Figure 10. UV luminescence of colloidal systems based on water and LaF_3. a: 1 – LaF_3, 2 – 4 – $LaF_3:Tb^{3+}$. b: 1 – LaF_3, 2 – 4 – $LaF_3:Eu^{3+}$. Phosphor concentration, $mg \cdot mL^{-1}$: 1 – 0.01; 2 – 0.02; 3 – 0.05; 4 – 0.1.

Figure 11 shows the UV luminescence spectrum of a sample of nanodispersed $LaF_3:Tb^{3+}$ when diluted in water at a concentration of 0.5 mg $\cdot mL^{-1}$ (curve 1). It's seen that the bands are present in spectrum, which are characteristic for the structure of $LaF_3:Tb^{3+}$ (Liu, Chen, Wang et al., 2008). In Figure 11, for comparison, the luminescence spectrum is shown for a non-activated LaF_3 sample under the same conditions (curve 2).

Figure 12 shows a typical X-ray luminescence spectrum of nanodispersed samples of $LaF_3:Tb^{3+}$. It's seen that 4 characteristic bands of X-ray luminescence are observed in the spectrum with maxima at 489.9, 542.7, 584.5, 621.7 nm, and when excited by UV and X-rays, their positions coincide well, which may indicate participation of the same centers in the optical electronic transitions.

Figure 13 shows the UV luminescence spectrum of a sample of nanodispersed $LaPO_4:Tb^{3+}$ when diluted in water at a concentration of 0.5 $mg \cdot mL^{-1}$ (curve 1). Curve 2 corresponds to the luminescence spectrum of $LaPO_4$ sample at a dilution of 0.05 mg/mL. It's seen that there are 4 main bands in the spectrum with maxima at 488, 543, 584, 622 nm, which can be associated with the corresponding electronic transitions involving Tb^{3+} energy levels in $LaPO_4$ band gap.

Figure 14 shows a typical X-ray luminescence spectrum of nanodispersed samples of $LaPO_4:Tb^{3+}$. It's seen that in the spectrum there are also 4 characteristic bands of X-ray luminescence, and the positions of their maxima are in good agreement with the corresponding of $LaF_3:Tb^{3+}$ structures (Figure 12), when excited by UV or X-rays. This may indicate, in particular, that under the conditions used for the synthesis of nanosized particles, the difference in the dielectric properties of lanthanum fluoride and phosphate matrices has a little effect on the position of energy levels in their band gap when activated with terbium ions.

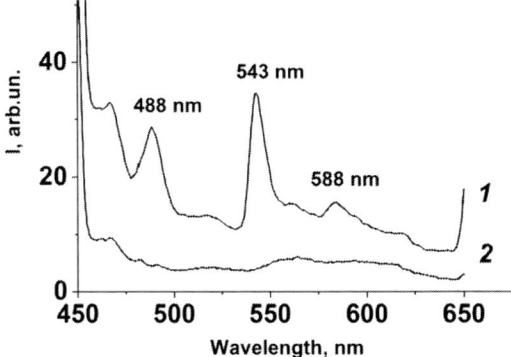

Figure 11. UV luminescence spectrum of samples of nanodispersed $LaF_3:Tb^{3+}$ when diluted in water at a concentration of 0.5 mg·mL^{-1} (curve 1) and non-terbium-activated LaF_3 (curve 2) under the same conditions. $T \sim 300$ K.

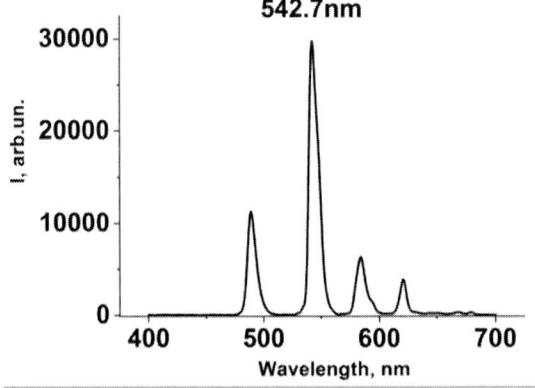

Figure 12. Typical X-ray luminescence spectrum of nanodispersed samples of $LaF_3:Tb^{3+}$. $T \sim 300$ K.

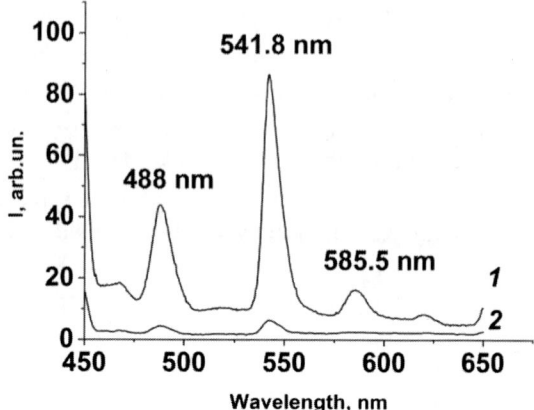

Figure 13. Spectrum of UV luminescence of samples of nanodispersed LaPO$_4$:Tb^{3+} when diluted in water at a concentration of 0.5 mg·mL^{-1} (curve 1) and 0.05 mg·mL^{-1} (curve 2). $T \sim 300$ K.

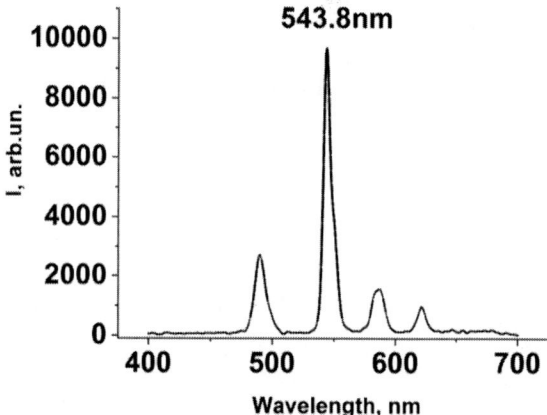

Figure 14. Typical X-ray luminescence spectrum of nanodispersed samples of LaPO$_4$:Tb^{3+}. $T \sim 300$ K.

The synthesized Fe$_3$O$_4$ NPs in the original ensemble were characterized by sizes of 3-23 nm and a single-domain state. An ensemble of magnetite NPs with an average size of 11 nm was used in this work. The specific surface area of the synthesized ensemble of magnetite was $S_{sp} = 105$ m^2·g^{-1}. Magnetite was characterized by coercive force $H_c = 55.0$ Oe, specific saturation magnetization $\sigma_s = 56.2$ Gs·cm^3·g^{-1}, relative residual magnetization $M_r/M_s = 0.2$, and it can be used as a magnetically controlled carrier for targeted drug delivery. For the magnetic fluid synthesized on the basis of water (physiological solution) and the studied ensemble of magnetite, the magnetization curve was without hysteresis, which indicated, in particular, the presence of superparamagnetic properties of Fe_3O_4 NPs (Abramov, Turanska, and Gorbyk, 2018a; Abramov, Turanska, and Gorbyk, 2018b) and lack of their aggregation in the magnetic fluid.

$Fe_3O_4/LaF_3:Tb^{3+}$ NC was obtained by precipitation of lanthanum fluoride on the surface of the original ensemble of Fe_3O_4 NPs, similar to the methods (Abramov, Turanska, and Gorbyk, 2018a; Abramov, Turanska, and Gorbyk, 2018b), which allow to obtain NC with a structure of core-shell type. The process of precipitation of $LaF_3:Tb^{3+}$ on the surface of single-domain Fe_3O_4 was performed according to the synthesis variant 1, described above.

Figure 15, *a* shows the TEM image of the ensemble of particles of NC $Fe_3O_4/LaF_3:Tb^{3+}$. It's seen that their shape is close to spherical, which is characteristic of Fe_3O_4 NPs and NC of core-shell type. X-ray diffraction patterns (Figure 15, *b*) confirm the presence of lanthanum fluoride in the structure of NC. It should be noted that the synthesis conditions of NC did not significantly change the magnetic properties of original Fe_3O_4. The X-ray luminescence spectrum of $Fe_3O_4/LaF_3:Tb^{3+}$ NC in the composition of the magnetic fluid is shown in Figure 15, *c*. The presence of a band with maximum at 543 nm may indicate participation of just $LaF_3:Tb^{3+}$ in X-ray luminescence, in this case its low intensity is due to the fairly small mass part of X-ray phosphor in the structure of NC. Note that just under such conditions, the magnetic properties of the ensemble of particles of NC $Fe_3O_4/LaF_3:Tb^{3+}$ best correspond to the properties of the original ensemble of NPs of single-domain Fe_3O_4, which acts as a magnetically controlled carrier. Further development of the above approaches to the creation of NC $Fe_3O_4/LaF_3:Tb^{3+}$ and their optimization may become the basis for the creation of new multifunctional magnetically controlled remedies of optopharmacology, in particular, for targeted delivery and local therapy of tumor diseases.

Samples of nanostructured composites based on 60S bioglass and lanthanum fluoride or phosphate activated by Tb^{3+} were synthesized. Aqueous suspensions (Figure 16, *a*) and dry samples (Figure 16, *b*) of $60S:LaF_3:Tb^{3+}$ (Kusyak, Petranovska, Turanska et al., 2021) and $60S:(LaPO_4:Tb^{3+})$ composites showed the presence of luminescence when excited by UV radiation.

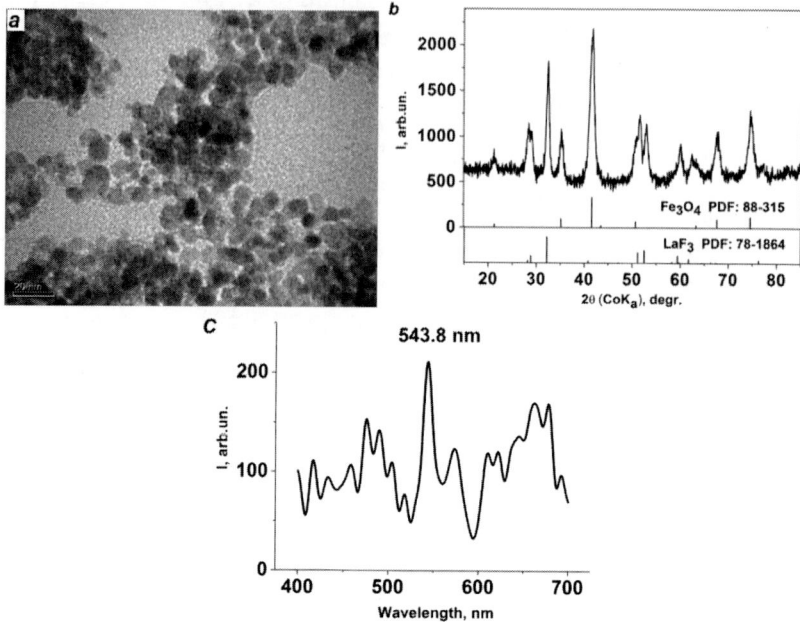

Figure 15. TEM images of NC $Fe_3O_4/LaF_3:Tb^{3+}$: scale of 20 nm (*a*), XRD patterns (*b*), X-ray luminescence spectrum (*c*) of samples of colloidal systems based on water and nanodispersed NC $Fe_3O_4/LaF_3:Tb^{3+}$ ($C = 0.1$ mg mL^{-1}). $T \sim 300$ K.

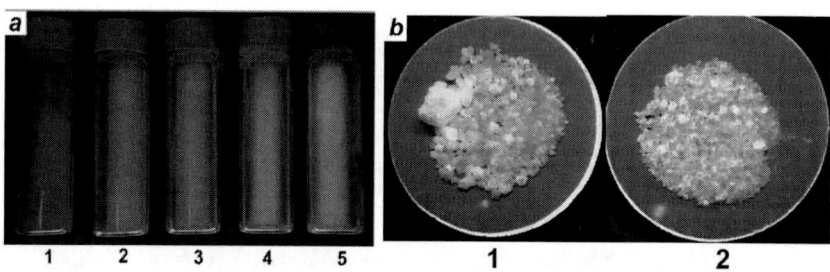

Figure 16. UV luminescence: *a* – colloidal systems based on water, 60S BG and LaF$_3$. 1 – H$_2$O, 60S BG (0.1 mg mL^{-1}), 2-5 – H$_2$O, 60S BG (0.1 mg mL^{-1}), contents of LaF$_3$:Tb^{3+} (mg mL^{-1}): 2 – 0.01, 3 – 0.05, 4 – 0.075, 5 – 0.1. All samples were sonicated for 5 minutes; *b* – samples of dispersed composites 60S:(LaPO$_4$:Tb^{3+}), synthesized at 700 (1) and 900 (2) °C.

Figures 17, *a*, *b* show the X-ray luminescence spectra of nanostructured composites 60S:(LaPO$_4$:Tb^{3+}) synthesized at 700 and 900°C, respectively. The presence of a band with a maximum of 545 nm may indicate that X-ray luminescence is associated with energy levels of activator Tb^{3+}. According to experimental data, an increase in the temperature of synthesis of composite 60S:($LaPO_4$:Tb^{3+}) to 900°C promotes an increase in the intensity of X-ray luminescence of the composite in the bands with maxima at 545 and 488 nm.

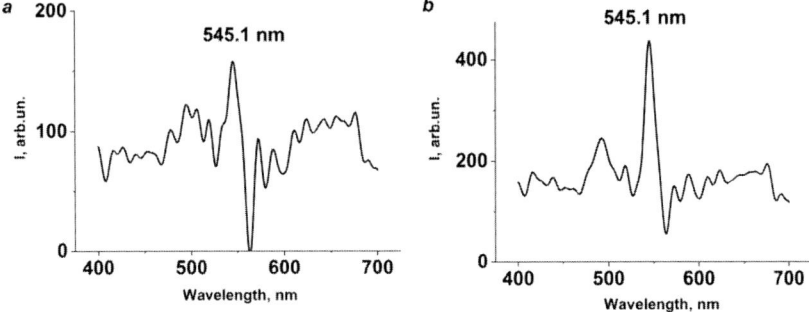

Figure 17. X-ray luminescence of samples of 60S:($LaPO_4$:Tb^{3+}) composites synthesized at 700 (*a*) and 900°C (*b*).

These data indicate the prospects of research into nanodispersed phosphors based on lanthanum fluoride and phosphate, their composites with magnetically sensitive carriers and bioactive glass, for use in optopharmacology and photodynamic therapy of tumors localized in cranial organs and bone tissues.

Conclusion

Terbium-activated nanocrystalline lanthanum fluoride and lanthanum phosphate of hexagonal syngony were synthesized. Structural properties, chemical activity of surface, UV and X-ray luminescence spectra of the synthesized crystals have been studied. The possibility is shown to use them in nanocomposites with magnetically sensitive nanosized drug carriers and bioactive sol-gel glass.

The acid-base nature of the active surface centers of LaF_3 and LaF_3:Tb^{3+} NPs was determined. The parts of negatively $\alpha-$, positively α^+ charged and neutral α^0 active centers were calculated in the pH range of 2.4 – 12.7. The detected active centers of the surface can be represented by acidic (La^{3+}) and base (F^-) Lewis centers, as well as base Bronsted centers (OH^- groups). The obtained data are useful for optimization of the conditions of adsorption immobilization of molecules of photosensitizers from physiological solution onto the surface of phosphors based on lanthanum fluoride.

Ensembles of particles of magnetically sensitive NC $Fe_3O_4/LaF_3:Tb^{3+}$ of the core-shell type were synthesized. Conditions for the synthesis of NC did not significantly change the magnetic properties of their cores – the original single-domain Fe_3O_4 NPs. 60S BG composites with nanodispersed crystalline $LaF_3:Tb^{3+}$ and $LaPO_4:Tb^{3+}$ in the dry state, and in distilled water, showed the presence of luminescence upon excitation by UV and X-rays.

These data indicate the potential of research into nanodispersed phosphors based on lanthanum fluoride and lanthanum phosphate, their composites with magnetically sensitive nanosized carriers and bioactive glass, for use in optopharmacology and photodynamic therapy of tumor diseases localized in cranial organs and bone tissues. In addition, the results of research can be useful for technical applications, in particular, in the creation of luminescent detectors of high-energy electromagnetic radiation, the development of photo- and optoelectronic devices etc.

References

Abramov, M. V., Kusyak, A. P., Kaminskiy, O. M., Turanska, S. P., Petranovska, A. L., Kusyak, N. V. and Gorbyk P. P. (2017). Magnetosensitive Nanocomposites Based on Cisplatin and Doxorubicin for Application in Oncology. In *Horizons in World Physics*. 293:1-56.

Abramov, M. V., Turanska, S. P. and Gorbyk, P. P. (2018a). Magnetic properties of nanocomposites of a superparamagnetic core–shell type. *Metallofiz. Noveishie Technol.*, 40(4):423-500 (in Ukrainian).

Abramov, M. V., Turanska, S. P. and Gorbyk, P. P. (2018b). Magnetic Properties of Fluids Based on Polyfunctional Nanocomposites of Superparamagnetic Core–Multilevel Shell Type. *Metallofiz. Noveishie Technol.*, 40(10):1283-1348 (in Ukrainian).

Buryanov, A. A., Chornyi, V. S., Dedukh, N. V., Dubok, V. A., Protsenko, V. V., Omelchenko, T. N., Vakulich, M. V., Lyanskorunskiy, V. N., Shapovalov, V. S. and Abudeikh, U. (2019). Peculiarities of regenerative reactions in filling bone defects with bioglass in combination with autologous plasma enriched with platelets. *Trauma*, 20(6):56-61 (in Russian).

Buryanov, O. A., Chornyi, V. S., Protsenko, V. V., Shapovalov, V. S. and Kusyak, V. A. (2018). Analysis of replacement of bone defects by calcium phosphate biomaterials in bone diseases. *Litopys Travmat. Ortoped.*, 1-2(37-38):111-116 (in Ukrainian).

DiMaio, J., Kokuoz, B., James, T. L., Harkey, T., Monofsky, D. and Ballato, J. (2008). Photoluminescent characterization of atomic diffusion in core-shell nanoparticles. *Opt. Exp.*, 16(16):11769-11775.

Dutra, C. E. A., Pereira, M. M., Serakides, R. and Rezende, C. M. F. (2008). In vivo evaluation of bioactive glass foams associated with platelet-rich plasma in bone defects. *J. Tissue Eng. Regen. Med.,* 2(4):221-227.

Foster, K. A., Oster, C. G, Mayer, M. M., Avery, M. L. and Audus, K. L. (1998). Characterization of the A549 cell line as a type II pulmonary epithelial cell model for drug metabolism. *Exp. Cell Res.,* 243(2):359-366.

Gorbyk, P. P. (2020). Biomedical nanocomposites with nanorobot functions: state of research, development, and prospects of practical introduction. *Him. Fiz. Tehnol. Poverhni*, 11(1):128-143 (in Ukrainian).

Gorbyk, P. P., Lerman, L. B., Petranovska, A. L., Turanska, S. P. and Pylypchuk, I. V. (2016). Magnetosensitive Nanocomposites with Hierarchical Nanoarchitecture as Biomedical Nanorobots: Synthesis, Properties, and Application. In *Fabrication and Self-Assembly of Nanobiomaterials, Applications of Nanobiomaterials*. Elsevier. 289-334.

Gorobets', S. V., Gorobets', O. Y., Gorbyk, P. P. and Uvarova, I. V. (2018). *Functional Bio- and Nanomaterials of Medical Destination*. Kyiv: Kondor (in Ukrainian).

He, H., Xie, M. Y., Ding, Y. and Yu, X. F. (2009). Synthesis of Fe3O4@LaF3:Ce,Tb nanocomposites with bright fluorescence and strong magnetism. *Applied Surface Science*, 255(8):4623-4626.

Hench, L. L. and Fielder, E. (2004). Biological Gel-Glasses. In *Sol-Gel Technologies for Glass Producers and Users,* eds. M. A. Aegerter and M. Mennig. Boston: Springer.

Hsiu-Wen, C., Chien-Hao, H., Chien-Hsin, Y. and Tzong-Liu, W. (2020). Synthesis, optical properties, and sensing applications of $LaF_3:Yb^{3+}/Er^{3+}/Ho^{3+}/Tm^{3+}$ upconversion nanoparticles. *Nanomater.,* 10:2477-2498.

Jing, K., Guo, X., Diao, X., Wu, Q., Jiang, Y., Sun, Y. and Zhu, Y. (2015). Synthesis and characterization of dipicolinate sensitized $LaF_3:Tb^{3+}$ nanoparticles and their interaction with bovine serum albumin. *J. Lumin.,* 157:184-192.

Kasturi, S., Marikumar, R. and Vaidyanathan, S. (2018). Trivalent rare-earth activated hexagonal lanthanum fluoride ($LaF_3:RE^{3+}$, where RE=Tb, Sm, Dy and Tm) nanocrystals: Synthesis and optical properties. *Luminescence*, 33(5):897-906.

Kemnitz, E. and Coman, S. (2016). Nanoscaled Metal Fluorides in Heterogeneous Catalysis. In *New Materials for Catalytic Applications*. Elsevier. 133-191.

Kusyak, A. P., Petranovska, A. L., Dubok, V. A., Chornyi, V. S., Bur'yanov, O. A., Korniichuk, N. M. and Gorbyk, P. P. (2021). Adsorption immobilization of chemotherapeutic drug cisplatin on the surface of sol-gel bioglass 60S. *Functional Materials,* 28(1):97-105.

Kusyak, A. P., Petranovska, A. L., Turanska, S. P., Oranska, O. I., Shuba, Y. M., Kravchuk, D. I., Kravchuk, L. I., Chornyi, V. S., Bur'yanov, O. A., Sobolevs'kyy, Y. L., Dubok, V. A. and Gorbyk, P. P. (2021). Synthesis and properties of nanostructures based on lanthanum fluoride for photodynamic therapy of tumors of the cranial cavity and bone tissue. *Him. Fiz. Tehnol. Poverhni*, 12(3):216-225.

Liu, Y., Chen, W., Wang, S., Joly, A. G., Westcott, S. and Woo, B. K. (2008). X-ray luminescence of $LaF_3:Tb^{3+}$ and $LaF_3:Ce^{3+}$, Tb^{3+} water-soluble nanoparticles. *J. Appl. Phys.,* 103(6):063105. https://doi.org/10.1063/1.2890148.

Mangaiyarkarasi, R., Chinnathambi, S., Karthikeyan, S., Aruna, P. and Ganesan, S. (2016). Paclitaxel conjugated Fe_3O_4@LaF_3:Ce^{3+},Tb^{3+} nanoparticles as bifunctional targeting carriers for cancer theranostics application. *J. Magn. Magn. Mater.*, 399:207-215.

Medkov, M. A., Steblevskaya, N. I. and Belobeletskaya, M. V. Patent of RF № 2617348 (24.04.2017) "Method for producing of lanthanum phosphate phosphor activated with cerium and terbium" (in Russian).

Min-Hua, C., Yi-Jhen, J., Sheng-Kai, W., Yo-Shen, C., Nobutaka, H. and Feng-Huei, L. (2017). Non-invasive photodynamic therapy in brain cancer by use of Tb^{3+}-doped LaF_3 nanoparticles in combination with photosensitizer through X-ray irradiation: a proof-of-concept study. *Nanoscale Res. Let.*, 12:62-68.

Patro, L. N., Kamala Bharathi, K. and Ravi Chandra Raju, N. (2014). Microstructural and ionic transport studies of hydrothermally synthesized lanthanum fluoride nanoparticles. *AIP Adv.*, 4:127139. doi:10.1063/1.4904949.

Roco, M. C., Williams, R. S. and Alivisatos, P. (2002). *Vision for Nanotechnology R&D in the Next Decade*. Dordrecht: Kluwer Acad. Publ.

Shpak, A. P. and Gorbyk, P. P., eds. (2009). *Nanomaterials and Supramolecular Structures: Physics, Chemistry, and Applications*. Netherlands: Springer. 63-78.

Tang, Y., Hu, J., Elmenoufy, A. H. and Yang, X. (2015). Highly efficient FRET system capable of deep photodynamic therapy established on X-ray excited mesoporous LaF_3:Tb scintillating nanoparticles. *ACS Appl. Mater. Int.*, 7(22):12261-12269.

Tressaud, A. (2010). *Functionalized Inorganic Fluorides: Synthesis, Characterization and Properties of Nanostructured Solids*. Chichester: John Wiley & Sons. 101-139.

Zhang, F., Braun, G. B., Pallaoro, A., Zhang, Y., Shi, Y., Cui, D., Moskovits, M., Zhao, D. and Stucky, G. D. (2011). Mesoporous multifunctional upconversion luminescent and magnetic "nanorattle" materials for targeted chemotherapy. *Nano Let.*, 12(1):61-67.

Chapter 4

Removal and Recovery of Lanthanum from Aqueous Solutions by Biosorption

Ellen C. Giese[*]
Center for Mineral Technology, CETEM, Rio de Janeiro, RJ, Brazil

Abstract

Biosorption is a cost-effective and simple technique for removing heavy metals and rare earth elements from an aqueous solution. Biosorption has also been considered a highly relevant green technology for replacing conventional unit operations of extractive metallurgy, viz. precipitation, liquid-liquid, solid-liquid extraction, and ion exchange. Biosorption is a physicochemical and metabolically-independent biological process based on various mechanisms, including absorption, adsorption, ion exchange, surface complexation, and precipitation, representing a biotechnological, cost-effective, innovative way for the recovery of lanthanum from aqueous solutions. This mini-review provides an overview and current scenario of biosorption technologies existing to recover lanthanum, seeking to address the possibilities of using a biotechnological approach for the recovery and separation of this high valued element in the REE production chain.

Keywords: lanthanum, biosorption, biosorbents

---

[*] Corresponding Author's Email: egiese@cetem.gov.br

In: What to Know about Lanthanum
Editor: Catherine C. Bradley
ISBN: 979-8-88697-615-1
© 2023 Nova Science Publishers, Inc.

Introduction

Among the factors driving the growth of the rare earth market, the high demand from emerging economies and the dependence on green technology that includes lanthanides in its composition are superimposed (Coey, 2020; Ballinger et al., 2020; Giese, 2019a; Giese, 2021; 2022a). As a result, this global market is estimated at 161,354.65 tonnes for 2022, with a forecast compound annual growth rate of more than 4% between 2022 and 2027 (Mordor Intelligence, 2022). The high market value of rare earth elements has been due to the high cost of producing and obtaining high purity rare earth oxides, in addition to inconsistent supply due to the Asia-Pacific market monopoly (Klinger, 2018; Souza et al., 2019; Giese, 2022a, 2022b).

The usual processes for separating rare earth elements involve classical methods of extractive hydrometallurgy, such as several steps of solvent extraction, use of ion-exchange resins, and selective precipitation (Croft et al., 2018; Asadollahzadeh et al., 2020; Jyothi et al., 2020; Panda et al., 2021). Ion exchange is preferred in the final stages of polishing the products obtained in the previous steps due to the low concentration of rare earth to be removed from the solution (Texier et al., 2002). In this context, biosorption has been described in the scientific literature as a potential ion exchange tool for the recovery and separation of lanthanum and other rare earth elements (Giese, 2022a; Giese et al., 2020; Heidelmann et al., 2022; Giese, 2022c).

The use of biosorption within an industrial process is encouraged by the lower operating cost and the biosorbent material. Among other advantages, we highlight the high efficiency in removing the element of interest from solutions containing low concentrations of the same, the easy regeneration of the biosorbent material, and the potential for recovery of the biosorbed aspects, in addition to the minimization of residues and fast adsorption kinetics and desorption (Giese, 2019b, 2020a; 2022b).

Biosorption has been studied as an alternative in extractive metallurgy processes for the recovery and concentration of high demand and high added value metals, such as gold, silver, and uranium, as well as rare earth elements. In this sense, the use of biosorbents is promising, as it also presents as a viable, cost-effective industrial process at the expense of the environmental impact of similar technologies (Korenevsky et al., 1999; Vijayaraghavan; Yun, 2008; Zoubolis et al., 2004).

Biosorption of lanthanum utilizing fungal, algal, or bacterial biomass (living, dormant, and dead cells) has been acknowledged as a possible alternative to the existing standard procedures for the treatment of industrial wastewater as well as for separation of this rare-earth element. The significant benefits of the biosorption process over conventional treatment methods include the following: low price, increased metal elimination, regeneration of biosorbent, and potential for metal recovery (Iftekhar et al., 2018).

This review article aims to investigate the trend of lanthanum recovery by biosorption based on the analysis of articles published in recent years.

Biosorbents

Biosorbent materials are porous substances that have a high surface area. Among the most commercially used adsorbent materials, activated carbon, zeolites, silica gel, and alumina stand out due to their high surface areas (>1000 m^2/g). In the case of activated carbon, the element recovery mechanism consists of physical adsorption. The physical forces that pure carbon exerts on impurities as classified as Van der Walls forces (Iqbal and Ashiq, 2007).

The biosorption of rare earth elements and heavy metals can be considered an ion-exchange process based on the functional groups present in the coatings of organisms. Substances present in cell coatings, such as polysaccharides, glycoproteins, and lipids, mainly can act as functional groups in the uptake of rare earth elements through binding sites such as carboxyl, and amine, sulfhydryl, phosphate, and hydroxyl radicals, among others (Giese, 2020a).

The different functional groups present themselves as binding sites and are responsible for the adsorption capacity of lanthanum ions. Phosphates act as monodentate ligands with lanthanides and are found mainly in the form of N-acetylglucosamine in extracellular polymers or the cell wall of gram-positive bacteria (Ngwenya et al., 2009).

The main biosorbents used in studies related to equilibrium, kinetics, and thermodynamics of lanthanum adsorption include *Sargassum fluitans* (Palmieri et al., 2001), *Pinus brutia* (Kütahyali et al., 2010), *Platanus orientalis* (Sert et al., 2008), *Agrobacterium* sp. (Xu et al., 2011), shells (Vijayaraghavan et al., 2009) and bamboo (Chen, 2010), among others.

The search for alternatives to conventional methods that have low cost and high efficiency has boosted, in recent years, research on the use of different biosorbents in adsorption systems (Giese, 2020a, 2020b; Giese et al., 2020). Biosorbent encompasses all biomass, whether active (with metabolic activity) or inactive (without metabolic activity).

All biosorbents come from some biological form, such as vegetables, crustaceans, microorganisms, and animals. The mechanisms that occur during the retention of metal ions in a biosorbent are directly related to the chemical functional groups in the material (Giese, 2022a). These macromolecules have chemical functional groups such as alcohols, aldehydes, ketones, carboxylic acids, phenols, and ethers. These groups donate electrons to the metal cation, thus establishing the bond between the metal ion and the biosorbent (Demirbas, 2008).

To prepare a biosorbent, physical treatments must be carried out at a low cost, such as washing, drying, grinding, and sieving. The sequence of these procedures produces a granulated material which can then be used as an adsorbent. After this biosorbent preparation step, it can be used as an adsorbent of pollutants in treatment systems; therefore, a complete characterization of this material must be done, in addition to elucidating kinetic and isothermal adsorption parameters. Among the various parameters that define whether a biosorbent has applicability or not is its adsorption capacity.

The adsorption capacity of a biosorbent is given by (q) and represents the mass of adsorbate (in milligrams) retained by the mass of biosorbent (in grams). The adsorption capacity is affected by several operational factors, such as:

- contact time is when the contact between the biosorbent and the adsorbate occurs. The contact between them is optimized through the use of agitation systems. Adsorption is maximum when the system reaches equilibrium when the adsorbate concentration becomes constant in the solution.
- pH of the solution: the pH of the adsorption medium has its essential based on the changes provided both in the biosorbent and in the metal ion. In the biosorbent, this influence can activate or deactivate active sites, modifying the way they present themselves to the metal. In the case of metal ions, the increase in pH can form insoluble metal hydroxides that do not react with the biosorbent.
- mass of biosorbent: the amount of mass of biosorbent used influences the number of active sites present but also provides other phenomena such as the association between adsorbent particles, reducing the number of adsorption points.
- initial concentration of metal ions: this variable is crucial, as the difference in this concentration about the concentration of metal ions

in the vicinity of the biosorbent surface generates a driving force of diffusion that helps to drive ions to the adsorption sites.

Both Langmuir and Freundlich's adsorption isotherms can represent the relationship between the amount of metal adsorbed by the adsorbent and the unadsorbed component in solution at a constant temperature. These isotherms provide the equilibrium data necessary for the design of the adsorption system (Iftekhar et al., 2018). Multiple types of structures can be used as biosorbent. Lanthanide binding to the biosorbent surface occurs via a passive mechanism involving stoichiometric interaction between the metal and the reactive chemical groups, followed by intracellular accumulation of metal due to the simultaneous effects of growth and surface biosorption. Several researchers have investigated removing La(III) from dilution solutions using a wide variety of microbial species, as described below.

Algae and Microalgae

Among the algae studied, the group most used by a variety of rare earth is the brown seaweeds, mainly those of the *Sargassacea* family, and due to their basic biochemical constitution: high concentrations of alginate and carboxylic groups in abundance, both capable of capturing cations present in solution (Davis et al., 2003).

These organisms are recognized for their ability to bioaccumulate high concentrations of rare earth compared to other marine and even terrestrial organisms (Goecke et al., 2015; Heidelmann et al., 2022). In addition, its voluminous biomass allows sufficient samples for bioaccumulation studies, making it unfeasible to perform this environmental analysis with microalgae. For these reasons, it has been speculated that these organisms may have the highest biosorption capacity for rare earth.

However, in laboratory studies for La biosorption, the green freshwater microalgae *C. reinharditii* was able to biosorb 1.03 mmol/g in 5h (Birungi and Chirwa, 2014), a value very close to the best result found with brown seaweed marine (*T. conoides*) of 1.11 mmol/g in 6h (Vijayaraghavan et al., 2010, 2011).

For lanthanum ions' adsorption to occur under algae's surface, it is necessary to displace pre-existing cations, with Ca being the main one. For example, Diniz and Voleski (2005) demonstrated that the exchange between Ca ions for binding with La, Eu, and Yb in *Sargassum polycystum* occurs in a

1:1 ratio; that is, the same concentration of Ca released in solution corresponded to the same concentration of soil are linked.

This cation exchange was also observed in the study by Vijayaraghavan et al., (2010) with the brown algae *Turbinaria conoides*, where energy scattering X-ray spectroscopy (ED-XRF) analysis showed that the Ca peaks on the cell surface were reduced when new peaks of La, Ce, Yb, and Eu biosorbents were present.

Birungi and Chirwa (2014) studied lanthanum ions' biosorption by four algae species (*D. multivariabilis, S. acuminutus, C. saccharophilum*, and *S. bacillaris*). The alga *C. saccharophilum* obtained more effective removal of lanthanum ion in a shorter time when it was in low concentration (15 mg/L), and in less than 15 min it was observed the removal of 94.34% of La. While at higher concentrations, 50 mg/L, and 150 mg/L, the alga *C. reinhardtii* was the one that most removed La, with rates of 38 mg/L and 75 mg/L, respectively, after 15-30 min of contact.

Some studies show that, in the mixture, the affinity of rare earth with functional groups will be related to its atomic mass, electronegativity, and ionic radius, which is higher when the element has lower atomic mass, lower electronegativity, and higher ionic radius (Vijayaraghavan and Balasubramanian, 2015).

For the alga *T. conoides*, the presence of more than one ion in the solution promoted competition for binding sites. The lanthanum ion biosorption rate, for example, decreased by 75.8% in the presence of Eu, Ce, and Yb when compared to their exposure, and the order of affinity observed was La > Ce > Eu > Yb (Vijayaraghavan et al., 2010). Coincidentally, this is the order of increasing atomic mass [La (138.9) < Ce (140.1) < Eu (151.9) < Yb (173.0)], increasing electronegativity [La(1,1) < Ce (1.12) = Eu (1.12) < Yb (1.21)] and decreasing ionic radius [(117.2) > Ce (115) > Eu (108.7) > Yb (100, 8)].

For other strains, Diniz and Voleski (2005 a, b) observed that in mixtures, the affinity of rare earth followed the order Eu > La > Yb. On the other hand, Oliveira et al., (2012) observed an increasing affinity following the order Gd < Nd < Pr < Sm < Eu < La. Thus, the affinity coefficient would not necessarily be related to the properties of rare earth, such as example, the ionic radius.

The maximum biosorption capacities for the brown alga *Cystoseira* were 185.44 mg/L for La and 172.33 mg/L for Ce(III), respectively. However, Mohammad et al., (2019) observed that the biosorption capacity of both metal ions decreased in the presence of Th(IV). Table 1 summarizes the studies of lanthanum biosorption by algae and the maximum biosorption capacity (q) per volume of biomass.

Table 1. Biosorption of lanthanum by algae

Specie	q	Reference
Sargassum fluitans	0.53 mmol/g	Palmieri and Garcia, 2011
Sargassum fluitans	0.61 mmol/g	Palmieri et al., 2001
Sargassum polycustum	1.0 mmol/g	Diniz; Voleski, 2005
Sargassum sp.	0.66 mmol/g	Oliveira; Garcia, 2009
Turbinaria conoides	1.11 mmol/g	Vijayaraghavan et al., 2010; 2011
Promidium	335 mg/Kg	Kim et al., 2011
Sargassum sp.	0.57 mmol/g	Oliveira et al., 2012
Sargassum sp.	0.23 mmol/g	Oliveira et al., 2012
Sargassum hemiphylum	0.70 mmol/g	Kano, 2013
Ulva pertusa	0.93 mmol/g	Kano, 2013
Schizymenia dubyi	0.65 mmol/g	Kano, 2013
Chloroidium saccharophilum	0.93 mmol/g	Birungi and Chirwa, 2013
Stichococcus bacillaris	0.36 mmol/g	Birungi and Chirwa, 2013
Desmodesmus multivariabilis	0.72 mmol/g	Birungi and Chirwa, 2013
Chlorella vulgaris	0.53 mmol/g	Birungi and Chirwa, 2013
Scenedesmus acuminutus	0.79 mmol/g	Birungi and Chirwa, 2013
Chlorella vulgaris	1.11 mg/g	Heidelmann et al., 2017
Cystoseira	185.44 mg/kg	Keshtar et al., 2019

Bacteria, Fungi, and Yeast

Among the bacterial strains studied as lanthanum biosorbents, those belonging to the *Bacillus* genera stand out (Takahashi et al., 2005; Tsuruta, 2007; Coimbra et al., 2017; Jordão and Giese, 2019), *Streptomyces* (Tsuruta, 2006a), *Pseudomonas* (Andrès et al., 2000; Philip et al., 2000; Texier et al., 2002), *Myxococcus* (Merroun et al., 2003).

Bacteria are widely studied for their potential as biosorbents because they have a high surface area about their small size, which facilitates the process of adsorption of metal ions in solution (Vijayaraghava and Yun, 2008).

The biosorption capacity, both for gram-positive bacteria such as *Bacillus subtilis* and gram-negative bacteria such as *Pseudomonas aeruginosa*, will depend on the composition of the bacterial cell wall; being necessary to evaluate the growth time and the composition of the culture medium to find the ideal biosorbent (Andrès et al., 2000).

The pretreatment of bacterial cells with solutions of NaCl, HCl, or NaOH may be necessary so that the active sites on the cell surface present sufficient negative charge density to favor the interaction between the rare earth and the biosorbent material.

For *B. subtilis*, for example, pretreatment with 1M NaOH favored the sorption process of La, equal to 100% after 40 minutes of contact with 100 µM La solution at pH 3.0. However, with the cells pretreated with 1M HCl, the maximum sorption was only 34% after 60 minutes (Jordão and Giese, 2019).

On the other hand, both caustic and acid pretreatments did not promote the increase of La and Ce biosorption by the *Agrobacterium* sp. HN1. In this case, the parameters pH, temperature, time, and age of the colony were the ones that influenced the biosorption process; the ideal condition is obtained in the presence of a solution containing La (15 mg/L), Ce (10 mg/L), 300 mg cells of 28 hours/L of the solution, pH 6.8, 30°C, 150 rpm and 2 h of contact (Xu et al., 2011).

The bacterium *B. subtilis* is considered a model for binding light and rare heavy earth in gram-positive bacteria when in an acidic solution. It is known that between pH 2.5 and 4.5, these bacteria have a specific binding site for rare-earth elements, which have low affinity for La, Ce, Pr, Nd (light) and higher affinity for Tm, Yb, Lu (heavy) (Martinez et al., 2014).

For the bacterium *Pseudomonas aeroginosa*, the order of selectivity of biosorption of La, Eu, and Yb ions were equal to Eu = Yb > La. The authors also observed that cell inactivation at 37 or 70 °C did not show differences in rare-earth biosorption patterns. In addition, the presence of sodium, potassium, calcium, chloride, nitrate, and sulfate ions did not affect the rates of biosorbed rare earth (Texier et al., 1999).

Takahashi et al., (2010) identified a rare-earth binding site on the cell surface of B. subtilis using the extended X-ray absorption fine structure spectroscopy (EXAFS) method and reported that rare earth is capable of binding to active sites on the cell surface through the formation of complexes with phosphate groups by phosphoester-like bonds. Furthermore, the authors observed that bonds with heavy rare earth (Gd to Lu) showed a higher coordination number with phosphate than bonds with light rare earth (La to Sm).

Two strains of *S. cerevisiae*, one wild and one mutant (rim20Δ) were evaluated for the biosorption capacity of La. DiCaprio et al., (2016) observed that ion-biomass interactions occurred through bonds with carboxylic, amino, and phosphate groups for both strains. The two processes reached an equilibrium between 10 and 20 minutes, with qmax of 70 mg/g at pH 4.0 for the wild strain and 80 mg/g at pH 6.0 for the mutant strain.

Most studies involving fungi and yeasts use rare earth bioaccumulation experiments to separate them. In bioaccumulation, rare earth is adsorbed during microbial growth, and in these cases, the presence of secondary metabolites will also influence the adsorption mechanism of these elements together with the cell wall, which will not present a fixed composition since it changes according to the growth stages (Cejpková et al., 2016).

Table 2 summarizes lanthanum biosorption studies by bacteria, fungi, and yeasts and the amount of biosorbed element or qmax per volume of biomass.

Table 2. Biosorption of lanthanum by bacteria, fungi, and yeasts

Specie	Type	q	Reference
Bacillus subtilis	bacteria	1.5 mg/g	Coimbra et al., 2017
Bacillus subtilis	bacteria	1.1 mg/g	Giese and Jordão, 2019
Bacillus subtilis	bacteria	6.0 mg/g	Coimbra et al., 2019
Pseudomonas aeruginosa	bacteria	397 µmol/g	Texier et al., 1999
Agrobacterium sp. HN1	bacteria	23.26 mg/g	Xu et al., 2011
Saccharomyces cerevisiae	yeast	70 mg/g	Di Caprio et al., 2016
Pleurotus ostreatus	fungi	54.54 mg/g	Hussien, 2014
Botryosphaeria rhodina	fungi	1.01 mg/g	Giese et al., 2019
Penicillium simplicissimum	fungi	7.84 mg/g	Bergsten-Torralba et al., 2021

Various Biosorbents

The intensity of the biosorption capacity depends both on the chemical and physical characteristics of the adsorbent, as well as on the properties of the adsorbate. A suitable adsorbent has many active sites available for interaction with the species of interest. Some biosorbents may have their surfaces chemically modified to increase the number of active sites and, consequently, increase the adsorption capacity of metal ions. The main modifications include esterification of carboxyl groups and phosphates, methylation of amino groups, and hydrolysis of carboxylate groups (Dermibas, 2008). In nature, there are many biosorbents that, in their natural state and properly used, provide similar or higher adsorption capacity values than those presented by chemically modified materials.

Table 3 summarizes lanthanum biosorption studies by different biosorbent materials and the maximum biosorption capacity (q) per biomass mass.

Table 3. Biosorption of lanthanum by various biosorbents

Biosorbent material	q	Reference
Crab shell	90.9	Butnariu et al., 2015
Crab shell	140.1	Vijayaraghavan et al., 2009
Neem sawdust	160.2	Das et al., 2014
Pinus brutia leaf powder	22.94	Kütahyali et al., 2010
Plantanus orientalis leaf powder	28.65	Sert et al., 2008
Grapefruit peel	171.2	Torab-Mostaedi et al., 2015
Tangerine peel	154.86	Torab-Mostaedi, 2013
Orange peel	125	Butnariu et al., 2015
Pineapple crown	100	Butnariu et al., 2015
Corn style	76.9	Butnariu et al., 2015
Bone powder	8.7	Varshini and Das, 2014
Bamboo Charcoal	215	Chen et al., 2010

Conclusion

Currently, the biggest challenge in the rare-earth industry is to separate and recover the elements from obtaining pure rare-earth compounds due to the high chemical similarity between the group elements. Biosorption combines biotechnology with extractive hydrometallurgy as a potential alternative for the concentration of rare-earth elements through interactions between rare earth and specific active sites present in the biosorbent.

In general, studies involving biosorption of rare earth by algae, bacteria, fungi, and yeasts are still at an early stage, with few species of organisms explored, and not all rare earth have been studied. Developing a bioprocess to compose the production chain of rare earth is something promising and innovative for the mineral sector. Alternative routes and clean technologies for the extraction and separation of rare earth are necessary and fit the current trends that seek better performance of hydrometallurgical facilities.

Acknowledgments

Ellen C. Giese thanks for financial support from National Council for Scientific and Technological Development (CNPq) - Brazil.

References

Andrès, Y., Thouand, G., Boualam, M. and Mergeay, M. (2000) Factors influencing the biosorption of gadolinium by micro-organisms and its mobilisation from sand. *Applied Microbiology and Biotechnology*, 54:262-267.

Asadollahzadeh, M., Torkaman, R. and Torab-Mostaedi, M. (2020) Study on the feasibility of using a pilot plant Scheibel extraction column for the extraction and separation of lanthanum and cerium from aqueous solution. *Korean Journal of Chemistry Engineering*, 37:322-331.

Ballinger, B., Schmeda-Lopez, D., Kefford, B., Parkinson, B., Stringer, M., Greig, C. and Smart, S. (2020) The vulnerability of electric-vehicle and wind-turbine supply chains to the supply of rare-earth elements in a 2-degree scenario. *Sustainable Production and Consumption*, 22:68-76.

Bergsten-Torralba, L. R., Nascimento, C. R. S., Buss, D. F. and Giese, E. C. (2021) Kinetics and equilibrium study for the biosorption of lanthanum by *Penicillium simplicissimum* INCQS 40,211. *3 Biotech* 11:460.

Birungi, Z. S. and Chirwa, E. M. N. (2013) Phytoremediation of lanthanum using algae from eutrophic freshwater sources. *Proceedings of the Water Environment Federation*, 19:357-366.

Birungi, Z. S. and Chirwa, E. M. N. (2014) The kinetics of uptake and recovery of lanthanum using freshwater algae as biosorbents: Comparative analysis. *Bioresource Technology*, 160:43-51.

Butnariu, M., Negrea, P., Lupa, L., Ciopec, M., Negrea, A., Pentea, M., Sarac, I. and Samfira, I. (2015) Remediation of rare earth element pollutants by sorption process using organic natural sorbents. *International Journal of Environmental Research Public Health* 12:11278-11287.

Cejpková, J., Gryndler, M., Hršelová, H., Kotrba, P., Řanda, Z., Synková, I. and Borovička, J. (2016) Bioaccumulation of heavy metals, metalloids, and chlorine in ectomycorrhizae from smelter-polluted area. *Environmental Pollution* 218:176-185.

Chen, Q. (2010) Study on the adsorption of lanthanum (III) from aqueous solution by bamboo charcoal. *Journal of Rare Earths* 28:125-131.

Coey, J. M. D. (2020) Perspective and prospects for rare earth permanent magnets. *Engineering* 6:119-131.

Coimbra, N. V., Gonçalves, F. S., Nascimento, M. and Giese, E. C. (2019) Study of adsorption isotherm models on rare earth elements biosorption for separation purposes. *International Scholarly and Scientific Research & Innovation* 13:200-203.

Coimbra, N. V., Nascimento, M. and Giese, E. C. (2017) Avaliação do uso de biomassa bacteriana imobilizada na biossorção de terras-raras leves e médias. *Holos* 6:136-146.

Croft, C. F., Almeida, M. I. G. S., Cattrall, R. W. and Kolev, S. D. (2018) Separation of lanthanum (III), gadolinium (III) and ytterbium (III) from sulfuric acid solutions by using a polymer inclusion membrane. *Journal of Membrane Science* 545:259-265.

Das, D., Varshini, C. J. S. and Das, N. (2014) Recovery of lanthanum (III) from aqueous solution using biosorbents of plant and animal origin: batch and column studies. *Mineral Engineering*, 69:40-56.

Davis, T. A., Volesky, B. and Mucci, A. (2003) A review of the biochemistry of heavy metal biosorption by brown algae. *Water Research* 37:4311-4330.

Dermibas, A. (2008) Heavy metal adsorption onto agro-based waste materials: A review. *Journal of Harzadous Materials* 157:220-229.

Di Caprio, F., Altimari, P., Zanni, E., Uccelletti, D., Toro, L. and Pagnanelli, F. (2016) Lanthanum biosorption by different *Saccharomyces cerevisiae* strains. *Chemical Engineering Transactions* 49:37-42.

Diniz, V. and Volesky, B. (2005a) Biosorption of La, Eu and Yb using *Sargassum* biomass. *Water Research* 39:239-247.

Diniz, V. and Volesky, B. (2005b) Effect of counterions on lanthanum biosorption by *Sargassum* polycystum. *Water Research*, v. 39, n. 11, p. 2229-2236, 2005b.

Giese, E. C. (2019a) Challenges of biohydrometallurgy in the circular economy. *Insights in Mining, Science and Technology* 1:123.

Giese, E. C. (2019b) Prospecção de tecnologias relacionadas ao processo de biossorção de metais. *Revista GEINTEC: Gestão, Inovação e Tecnologias* 9:5046-5057.

Giese, E. C. (2020a) Biosorption as green technology for the recovery and separation of rare earth elements. *World Journal of Microbiology and Biotechnology* 36:52.

Giese, E. C. (2020b) Mining applications of immobilized microbial cells in alginate matrix: an overview. *Revista Internacional de Contaminación Ambiental* 36:775-787.

Giese, E. C. (2021) Assessment of the potential for the use of rare earths in biomaterials through patent analysis. *RECIMA21 - Revista Científica Multidisciplinar* 2:e27590.

Giese, E. C. (2022a) E-waste mining and the transition toward a bio-based economy: The case of lamp phosphor powder. *MRS Energy & Sustainability* s43581.

Giese, E. C. (2022b) Strategic minerals: global challenges post-COVID-19. *The Extractive Industries and Society* 101113.

Giese, E. C. (2022c) Natural biosorbents for lanthanides recovery. *Industrial Biotechnology* 18:71-76.

Giese, E. C. and Jordão, C. S. (2019) Biosorption of lanthanum and samarium by chemically modified *Bacillus subtilis* free cells. *Applied Water Science* 9:182.

Giese, E. C., Barbosa, A. M. and Dekker, R. F. H. (2019) Biosorption of lanthanum and samarium by viable and autoclaved mycelium of *Botryosphaeria rhodina* MAMB-05. *Biotechnology Progress* 36:e2783.

Giese, E. C; Silva, D. D. V., Costa, A. F. M., Almeida, S. G. C. and Dussan, K. J. (2020) Immobilized microbial nanoparticles for biosorption. *Critical Reviews in Biotechnology* 40:653-666.

Goecke, F., Zachleder, V. and Vítová, M. (2015) Rare earth elements and algae: physiological effects, biorefinery and recycling. In: *Algal Biorefineries*. Springer International Publishing, p. 339-363.

Heidelmann, G. P., Egler, S. G. and Giese, E. C. (2022) Lanthanides biosorption by immobilized microalgae using statistical design. *International Journal of Chemistry Studies* 6:67-72.

Heidelmann, G. P., Roldao, T. M., Egler, S. G., Nascimento, M. and Giese, E. C. (2017) Uso de biomassa de microalga para biossorção de lantanídeos. *Holos* 6:170-179.

Hussien, S. S. (2014) Biosorption of lanthanum on *Pleurotus ostreatus* basidiocarp. *International Journal of Biomedical Research* 26-36.

Iftekhar, S., Ramasamy, D. L., Srivastava, V. and Asif, M. B. Silanpää. (2018) Understanding the factors affecting the adsorption of lanthanum using different adsorbents: A critical review. *Chemosphere* 204:413-430.

Iqbal, M. J. and Ashiq, M. N. (2007) Adsorption of dyes from aqueous solutions on activated charcoal. *Journal of Hazardous Materials* 139:57-66.

Jyothi, R. K., Thenepalli, T., Ahn, J. W., Parhi, P. K., Chung, K. W. and Lee, J-Y. (2020) Review of rare earth elements recovery from secondary resources for clean energy technologies: Grand opportunities to create wealth from waste. *Journal of Cleaner Production* 267:122048.

Kano, N. (2013) Biosorption of lanthanides using select marine biomass. Engineering "Biomass Now - Sustainable Growth and Use." *Rijeka: Intech Open,* p. 101-124.

Kim, J. A., Dodbiba, G., Tanimura, Y., Mitsuhashi, K., Fukuda, N., Okaya, K., Matsuo, S. and Fujita, T. (2011) Leaching of rare-earth elements and their adsorption by using blue-green algae. *Materials Transactions* 52:1799-1806.

Klinger, J. M. (2018) Rare earth elements: Development, sustainability and policy issues. *The Extractive Industries and Society* 5:1-7.

Korenevsky, A. A., Sorokin, V. V. and Karavaiko, G. I. (1999) Biosorption of rare earth elements. *Process Mettalurgy* 9:299-306.

Kütahyali, C., Sert, Ş., Cetinkaya, B., Inan, S. and Eral, M. (2010) Factors affecting lanthanum and cerium biosorption on *Pinus brutia* leaf powder. *Separation Science and Technology* 45:1456-1462.

Martinez, R. E., Pourret, O. and Takahashi, Y. (2014) Modeling of rare earth element sorption to the Gram-positive *Bacillus subtilis* bacteria surface. *Journal of Colloid and Interface Science* 413:106-111.

Merroun, M. L., Ben Chekroun, K., Arias, J. M. and Gonzalez-Muñoz, M. T. (2003) Lanthanum fixation by *Myxococcus xanthus*: cellular location and extracellular polysaccharide observation. *Chemosphere* 52:113-120.

Mordor Intelligence. *Rare earth elements market - growth, trends, Covid-19 impact, and forecasts (2022-2027).* https://www.mordorintelligence.com/industry-reports/rare-earth-elements-market.

Ngwenya, B., Mosselmans, J. F. W., Magennis, M., Atkinson, K. D., Tourney, J., Olive, V. and Ellam, R. M. (2009) Macroscopic and spectroscopic analysis of lanthanide adsorption to bacterial cells. Geochimica et Cosmochimica Acta 73:3134-3147.

Oliveira, R. C. and Garcia Jr, O. (2009) Study of biosorption of rare earth metals (La, Nd, Eu, Gd) by *Sargassum* sp. biomass in batch systems: physicochemical evaluation of kinetics and adsorption models. *Advanced Materials Research* 71:605-608.

Oliveira, R. C., Guibal, E. and Garcia, O. (2012) Biosorption and desorption of lanthanum (III) and neodymium (III) in fixed-bed columns with *Sargassum* sp.: Perspectives for separation of rare earth metals. *Biotechnology Progress* 28:715-722.

Palmieri, M. C. and Garcia, O. (2011) Biosorption of erbium and ytterbium using biomass of microorganisms. Process Metallurgy, v. 11, p. 137-144, 2011.

Palmieri, M. C., Volesky, B. and Garcia, O. (2001) Biosorption of lanthanum using *Sargassum* fluitans in batch system. *Hydrometallurgy* 67:31-36.

Philip, L., Iyengar, L. and Venkobachar, C. (2000) Biosorption of U, La, Pr, Nd, Eu and Dy by *Pseudonomas aeruginosa*. *Journal of Industrial Microbiology and Biotechnology* 25:1-7.

Sert, Ş., Kütahyali, C., İnan, S., Talip, Z., Çetinkaya, B. and Eral, M. (2008) Biosorption of lanthanum and cerium from aqueous solutions by *Platanus orientalis* leaf powder. *Hydrometallurgy* 90:13-18.

Souza, A. C. S. P., Nascimento, M. and Giese, E. C. (2019) Desafios para a extração sustentável de minérios portadores de terras raras. *Holos* 1:1-9.

Takahashi, Y., Chatellier, T. X., Hattori, K. H., Kato, K. and Fortin, D. (2005) Adsorption of rare earth elements onto bacterial cell walls and its implication for REE sorption onto natural microbial mats. *Chemical Geology* 219:53-67.

Takahashi, Y., Yamamoto, M., Yamamoto, Y. and Tanaka, K. (2010) EXAFS study on the cause of enrichment of heavy REEs on bacterial cell surfaces. *Geochimica et Cosmochimica Acta* 74:5443-5462.

Texier, A. C., Andrès, Y., Faur-Brasquet, C. and Le Cloircç, P. (1999) Selective biosorption of lanthanide (La, Eu, Yb) ions by *Pseudomonas aeruginosa*. *Environmental Science and Technology* 33:489-495.

Texier, A. C., Andrès, Y., Faur-Brasquet, C. and Le Cloireç, P. Fixed-bed study for lanthanide (La, Eu, Yb) ions removal from aqueous solutions by immobilized *Pseudonomas aeruginosa*: experimental data and modelization. *Chemosphere*, v. 47, n. 3, p. 333-342, 2002.

Torab-Mostaedi, M. (2013) Biosorption of lanthanum and cerium from aqueous solutions using tangerine (*Citrus reticulata*) peel: equilibrium, kinetic and thermodynamic studies. *The Chemical Industry & Chemical Engineering Quarterly* 19:79-88.

Torab-Mostaedi, M., Asadollahzadeh, M. and Hemmati, A. (2015) Biosorption of lanthanum and cerium from aqueous solutions by grapefruit peel: equilibrium, kinetic and thermodynamic studies. *Research on Chemical Intermediates* 41:559-573.

Tsuruta, T. (2006) Bioaccumulation of uranium and thorium from the solution containing both elements using various microorganisms. *Journal of Alloys and Compounds*, 408-412:1312-1315.

Tsuruta, T. (2007) Accumulation of rare earth elements in various microorganisms. *Journal of Rare Earths* 25:526-532.

Varshini, C. J. S. and Das, N. (2014) Relevant approach to assess the performance of Biowaste materials for the recovery of lanthanum (III) from aqueous medium. *Research Journal of Pharmaceutical, Biological and Chemical Sciences* 5:88-94.

Vijayaraghavan, K. and Balasubramanian, R. (2015) Is biosorption suitable for decontamination of metal-bearing wastewaters? A critical review on the state-of-the-art of biosorption processes and future directions. *Journal of Environmental Management* 160:283-296.

Vijayaraghavan, K. and Yun, Y. S. (2008) Bacterial biosorbents and biosorption. *Biotechnology Advances* 26:266-291.

Vijayaraghavan, K., Mahadevan, A. Joshi, U. M. and Balasubramanian, R. (2009) An examination of the uptake of lanthanum from aqueous solution by crab shell particles. *Chemical Engineering Journal* 152:116-121.

Vijayaraghavan, K., Sathishkumar, M. and Balasubramanian, R. (2010) Biosorption of lanthanum, cerium, europium, and ytterbium by a brown marine alga, *Turbinaria conoides*. *Industrial & Engineering Chemistry Research* 49:4405-4411.

Vijayaraghavan, K., Sathishkumar, M. and Balasubramanian, R. (2011) Interaction of rare earth elements with a brown marine alga in multi-component solutions. *Desalination* 265:54-59.

Xu, S., Zhang, S., Chen, K., Han, J., Liu, H. and Wu, K. (2011) Biosorption of La^{3+} and Ce^{3+} by *Agrobacterium* sp. HN1. *Journal of Rare Earths* 29:265-270.

Biographical Sketch

Ellen Cristine Giese, PhD

Affiliation: Center for Mineral Technology (CETEM).

Education: Chemistry PhD

Business Address: Av Pedro Calmon Viana 900, Cidade Universitária, Rio de Janeiro, 21941890, Brazil

Research and Professional Experience:
Researcher at the Center for Mineral Technology (CETEM) at Extractive Metallurgy and Bioprocesses group since 2013, where coordinates projects in Hydrometallurgy, Biohydrometallurgy, Biotechnology and Urban Mining. Actual researcher fellow of productivity in technological development and innovative extension (CNPq-DT, Brazil) since 2017.

Bachelor's of Science (Chemistry) (2002), specialization in Applied Biochemistry (2004) and master's in Biotechnology (2005) at Universidade Estadual de Londrina. Doctorate in Food Science and Engineering (2008) at Universidade Estadual Paulista Júlio de Mesquita Filho (UNESP). Post-doctoral terms in Biotechnology projects at Universidad de Castilla-La Mancha (Spain, 2008-09), Lakehead University (Canada, 2009-10) and Universidade de São Paulo (2011-12).

Publications from the Last 3 Years:
Stanković, Martin, M., Goldmann, S., Gabler, H., Haubrich, F., Moutinho, V. F., Giese, E. C., Neumann, R., Stropper, J. L., Kaufhold, S., Ufer, K., Reiner, D., Marbler, H., Schippers, A. Effect of mineralogy on Co and Ni

extraction from Brazilian limonitic laterites via bioleaching and chemical leaching. *Minerals Engineering*, 107604, 2022.

Giese, E. C. Strategic Minerals: Global Challenges Post-COVID-19. *The Extractive Industries and Society*, 101113, 2022.

Giese, E. C. E-waste mining and the transition toward a bio-based economy: The case of lamp phosphor powder. *MRS Energy & Sustainability*, 2022.

Almeida, S. G. C.; Mello, G. F.; Santos, M. G.; Silva, D. D. V.; Giese, E. C.; Hassanpour, M.; Zhang, Z.; Dussan, K. J. Saccharification of acid-alkali pretreated sugarcane bagasse using immobilized enzymes from Phomopsis stipata. 3 *Biotech*, v. 12, p. 39, 2022.

Giese, E. C. Reflexões sobre os impactos do COVID-19 no setor mineral. *Recima* 21, V. 3, N. 4, P. E341332.

Souza, A. C. S. P.; Giese, E. C. Recovery of rare-earth elements from brazilian ion-adsorption clay: A preliminary study. *Orbital*, v.14, n. 1, p. 10-14, 2022.

Heidelmann, G. P.; Egler, S. G.; Giese, E. C. Lanthanides biosorption by immobilized microalgae using statistical design. *International Journal of Chemistry Studies*, v. 6, n. 1, p. 67-72, 2022.

Silva, L. M.; Giese, E. C.; Medeiros, G. A.; Fernandes, M. T.; Castro, J. A. Evaluation of the use of Burkholderia caribensis bacteria for the reduction of phosphorus content in iron ore particles. *Materials Research*, v. 25, p. e20210427, 2022.

Giese, E. C. Natural biosorbents for lanthanides recovery. *Industrial Biotechnology*, v. 18, p. 71-76, 2022.

Silva, L. M.; Giese, E. C.; Medeiros, G. A.; Fernandes, M. T.; Castro, J. A. Evaluation of the use of organic acids in the reduction of phosphorus content in iron ore particles. *SSRN Electronic Journal*, v. 1, p. 4019863, 2022.

Giese, E. C. Influence of organic acids on pentlandite bioleaching by Acidithiobacillus ferrooxidans LR. 3 *Biotech*, v. 11, p. 165, 2021.

Souza, A. C. S. P.; Giese, E. C. An exploratory study of recovery of rare-earth elements from monazite in mild conditions using statistical-mixture design. *International Research Journal of Multidisciplinary Technovation*, v. 3, p. 20-25, 2021.

Giese, E. C. Biomining in the post-COVID-19 circular bioeconomy: a "green dispute" for critical metals. *International Research Journal of Multidisciplinary Technovation*, v. 3, p. 35-38, 2021.

Giese, E. C. Avaliação do potencial do uso de terras-raras em biomateriais através da análise de patentes. *Revista Científica Multidisciplinar*

(Assessment of the potential for the use of rare earths in biomaterials through patent analysis. *Multidisciplinary Scientific Journal*), v. 2, p. e27590, 2021.

Bergsten-Torralba, L. R.; Nascimento, C. R. S.; Buss, D. F.; Giese, E. C. Kinetics and equilibrium study for the biosorption of lanthanum by Penicillium simplicissimum INCQS 40211. *3 Biotech,* v. 11, p. 460, 2021.

Giese, E. C.; Lins, F. A. F.; Xavier, L. H. Desafios da reciclagem de lixo eletrônico e as cooperativas de mineração urbana. *Brazilian Journal of Business*, v. 3, p. 3647-3660, 2021.

Xavier, L. H.; Duthie, A. C.; Giese, E. C.; Lins, F. A. F. Sustainability and the circular economy: A theoretical approach focused on e-waste urban mining. *Resources Policy*. 2021. v. 1, p. 101467.

Giese, E. C; Silva, D. D. V.; Costa, A. F. M.; Almeida, S. G. C.; Dussan, K. J. Immobilized microbial nanoparticles for biosorption. *Critical Reviews in Biotechnology*. 2020. v. 40, n. 5, p. 653-666.

Giese, E. C. Biosorption as green technology for the recovery and separation of rare earth elements. *World Journal of Microbiology and Biotechnology*. 2020. v. 36, n. 4, p. 52.

Giese, E. C. Mining applications of immobilized microbial cells in alginate matrix: an overview. *Revista Internacional de Contaminación Ambiental*. 2020. v. 36, n. 4, p. 775-787.

Bergsten-Torralba, L. R.; Magalhães, D. P.; Giese, E. C.; Nascimento, C. R. S.; Pinho, J. V. A.; Buss, D. F. Toxicity of three rare earth elements, and their combinations to algae, microcrustaceans, and fungi. *Ecotoxicology and Environmental Safety*. 2020. v. 201, p. 110795.

Giese, E. C. Challenges of biohydrometallurgy in the circular economy. *Insights in Mining, Science and Technology*, 2019. v. 1, n. 3, p. 123.

Giese, E. C; Barbosa, A. M.; Dekker, R. F. H. Biosorption of lanthanum and samarium by viable and autoclaved mycelium of Botryosphaeria rhodina MAMB-05. *Biotechnology Progress*. 2019. v. 36, p. e2783.

Souza, A. C. S. P.; Nascimento, M.; Giese, E. C. Desafios para a extração sustentável de minérios portadores de terras raras. *Holos* (Challenges for the sustainable extraction of rare earth ores. *Holes*). 2019. v. 1, p. 1-9.

Giese, E. C. Functionalized nano diamonds: Improving biomedical features using rare-earth elements. *Biomedical Journal of Scientific & Technical Research*. 2019. v. 22, p. 17035-17036.

Giese, E. C.; Jordao, C. S. Biosorption of lanthanum and samarium by chemically modified Bacillus subtilis free cells. *Applied Water Science*. 2019. v. 9, p. 182.

Giese, E. C. Inovações tecnológicas na biomineração de minérios lateríticos de níquel e cobalto. *Tecnologia Em Metalurgia, Materiais E Mineração* (Technological innovations in the biomining of nickel and cobalt lateritic ores. Technology in Metallurgy, Materials and Mining). 2019. v. 16, p. 1-5.

Giese, E. C. Prospecção de tecnologias relacionadas ao processo de biossorção de metais. *Revista Geintec: Gestão, Inovação E Tecnologias* (Prospecting technologies related to the metal biosorption process. Geintec Magazine: Management, Innovation and Technologies) 2019. v. 9, p. 5046-5057.

Coimbra, N. V.; Goncalves, F. S.; Nascimento, M.; Giese, E. C. Study of adsorption isotherm models on rare earth elements biosorption for separation purposes. *International Scholarly and Scientific Research & Innovation.* 2019. v. 13, p. 200-203.

Giese, E. C., Carpen, H. L., Bertolino, L. C., Schneider, C. L. Characterization and bioleaching of nickel laterite ore using Bacillus subtilis strain. *Biotechnology Progress.* 2019. p. e2860.

Giese, E. C. Evidences of EPS-iron (III) ions interactions on bioleaching process mini-review: The key to improve performance. *Orbital: The Electronic Journal Of Chemistry.* 2019. v. 11, p. 200-204.

Giese, E. C.; Xavier, L. H.; Lins, F. A. F. Biomineração urbana: o futuro da reciclagem de resíduos eletroeletrônicos. *Brasil Mineral* (São Paulo), v. 385, p. 36-39, 2018.

Giese, E. C. Rare earth elements: Therapeutic and diagnostic applications in modern medicine. *Clinical and Medical Reports*, v. 2, p. 1-2, 2018.

Index

A

antimicrobial, vii, 1, 2, 5, 11, 16, 17, 22, 25, 28, 29
antimicrobial activity(ies), vii, 2, 5, 11, 16, 17, 29
antioxidant activity(ies), 15, 16
aquatic environments, viii, 35, 37, 58, 60
aquatic organisms, v, vii, 35, 37, 38, 39, 40, 41, 48, 63

B

Bacillus subtilis, 101, 103, 106, 107, 111, 112
bacteria, 16, 97, 101, 102, 103, 104, 107, 110
binuclear complexes, 8, 9, 10, 22, 25, 32, 34
biosorbent materials, 96, 102, 103
biosorbents, 95, 96, 97, 98, 100, 101, 103, 104, 105, 106, 108, 110
biosorption, v, vii, ix, 95, 96, 97, 99, 100, 101, 102, 103, 104, 105, 106, 107, 108, 109, 110, 111, 112
bivalves, 41, 42, 49, 50

C

chemical nuclease(s), vii, 1, 2, 11
crustaceans, 42, 43, 51, 52, 98
crystal structure, 7, 18, 23, 25, 26, 28, 30, 31, 69, 77
crystalline, ix, 18, 66, 92
crystallization, 77
cytotoxic activities, vii, 1, 12, 13, 14, 15
cytotoxicity, 13, 27, 28, 48

D

DNA, 2, 11, 12, 13, 14, 15, 17, 22, 24, 26, 27, 28, 29, 32, 33, 34, 46, 48
DNA binder(s), 11

E

echinoderms, 42, 51
ecotoxicity, 36, 39, 40, 46, 48, 49, 55, 59, 62, 63
elimination, viii, 35, 36, 39, 40, 46, 50, 57, 58, 61, 64, 97

F

freshwater, viii, 35, 47, 52, 53, 55, 56, 60, 61, 99, 105
fungi, 16, 52, 60, 101, 103, 104, 111

G

gadolinium, 57, 61, 62, 64, 105
green algae, 53, 56, 60, 107

H

heat shock protein, 47, 50
heavy metals, ix, 40, 95, 97, 105

I

immobilization, ix, 66, 85, 92, 93
immobilized enzymes, 110
invertebrates, viii, 36, 41, 42, 43, 49
ions, 3, 5, 6, 8, 9, 10, 15, 16, 17, 18, 20, 31, 39, 47, 53, 54, 55, 68, 69, 70, 73, 74, 75,

77, 85, 87, 97, 98, 99, 100, 102, 108, 112

K

kinetics, 96, 97, 105, 107

L

La(III) complex, vii, 1, 3, 4, 5, 6, 7, 8, 9, 10, 11, 12, 13, 14, 15, 16, 17, 19, 20, 21, 22, 27, 29
La(III) ion, vii, 1, 3, 4, 5, 6, 8, 9, 10, 12, 15, 16, 17, 19, 20
Lanthanum (La), v, vii, viii, 1, 2, 3, 4, 5, 6, 7, 8, 9, 10, 11, 12, 13, 14, 15, 16, 17, 18, 19, 20, 21, 22, 23, 24, 25, 26, 27, 28, 29, 30, 31, 32, 33, 34, 35, 36, 37, 38, 39, 40, 41, 42, 43, 44, 45, 46, 47, 48, 49, 50, 51, 52, 53, 55, 56, 57, 58, 59, 60, 61, 62, 63, 64, 65, 69, 70, 71, 95, 99, 100, 102, 106, 107, 108, 109
lanthanum fluoride, v, vii, viii, 65, 66, 68, 70, 77, 82, 85, 87, 89, 91, 92, 93, 94
lanthanum phosphate, viii, 66, 78, 91, 92, 94
Lanthanum(III) complex, 2, 23, 24, 25, 26, 27, 29, 30, 32, 34
luminescent nanostructures, 66
luminescent properties, 17, 18, 19, 20, 30, 68

M

magnetite, 45, 56, 64, 66, 69, 70, 88
metal complexes, vii, 1, 2, 10, 11, 12, 13, 16, 17, 29
metal hydroxides, 98
metal ions, vii, 1, 2, 18, 19, 20, 61, 67, 98, 100, 101, 103
metals, 3, 39, 53, 54, 64, 96, 107, 110
microalgae, 52, 53, 55, 99, 106, 110

N

Na^+, 47
NaCl, 73, 75, 102
nanocomposites, viii, 66, 67, 69, 91, 92, 93
nanocrystals, 79, 86, 93
nanoparticles, ix, 16, 27, 28, 43, 53, 60, 66, 72, 81, 92, 93, 94, 106, 111
nanostructures, 66, 68, 84, 85, 93
naphthalene, 9, 25

O

optopharmacology, v, vii, viii, 65, 66, 68, 89, 91, 92
oxidative stress, 49, 54, 56

P

photodynamic therapy (PDT), v, vii, viii, 65, 66, 67, 68, 91, 92, 93, 94
photoluminescence, 19, 20, 29, 85
photoluminescence properties, 17, 19, 20, 29
photoluminescent, 2, 19, 24, 30, 92
photosensitizers, ix, 66, 67, 68, 92

Q

quercetin, 12, 14, 15, 26, 28

R

radiation, ix, 31, 66, 67, 68, 71, 76, 92
rare earth elements (REE), vii, ix, 35, 36, 37, 38, 39, 40, 44, 46, 47, 48, 51, 53, 54, 55, 57, 58, 59, 60, 61, 62, 63, 64, 85, 95, 96, 97, 105, 106, 107, 108, 109, 111, 112

S

Scandium (Sc), 36
sol-gel, viii, 66, 67, 71, 91, 93
sol-gel glass, viii, 66, 67, 71, 91

T

technology-critical elements (TCE), vii, 35
toxic effect, 38, 51, 52, 53, 54, 55, 58, 59, 64
toxicity, 38, 40, 46, 47, 48, 49, 50, 51, 53, 54, 55, 56, 58, 60, 61, 62, 63, 64
toxicology, 61, 63

U

UV radiation, 31, 85, 89

V

vertebrates, viii, 36, 41, 46

W

wastewater, 37, 97

X

X-ray diffraction, 3, 71, 77, 78, 89, 90

Y

yeast, 101, 103
Yttrium (Y), 6, 12, 18, 20, 22, 23, 24, 25, 26, 27, 28, 30, 31, 36, 44, 47, 49, 50, 51, 54, 61, 62, 63, 64, 93, 94, 105, 107, 108